REUNION TOUR

Donald Rechler
with
Margaret Fraser

Inquiries should be addressed to:
Biography for Everyone, LLC
951 Spanish Circle #444
Delray Beach, FL 33483
800-393-KATE (5283) or 954-461-5283

To obtain a copy of this book contact:
Donald and Judy Rechler
(FAX) 631-414-8402

Library of Congress Control Number: 2004109825

ISBN: 1-888069-18-X

Cover design by Caren Hackman

Printed in the United States of America

To Judy, my soul mate and life's partner, who fills my life with love.

To my family and friends . . . past, present and future.

NORTE
great-great grandmother

NORTE
great-great grandfather

NORTE (ROSETTE)
great grandmother

ROSETTE
great grandfather

-?- (OHRBACH)
great grandmother
"Little Grandmother"

OHRBACH
great grandfather

GOLDSTEIN
great grandmother

GOLDSTEIN
great grandfather

RECHLER
great grandmother

RECHLER
great grandfather

MILDRED ROSETTE (OHRBACH)
grandmother (maternal)

BENJAMIN OHRBACH
grandfather (maternal)

YETTA GOLDSTEIN (RECHLER)
grandmother (paternal)

MICHAEL RECHLER
grandfather (paternal)

GLORIA OHRBACH (RECHLER)
mother

WILLIAM RECHLER
father

DONALD RECHLER
1934

ROGER RECHLER
1942

iv

top: Yetta and Michael Rechler, my grandparents, looking very elegant, this picture, probably in Romania in the late 1800's.

bottom: Willie Rechler, Brother Danny and Sister Ruth (my Aunt Ruth), 1914.

top: Mildred Rosette (Ohrbach), my grandmother, Harlem hat designer, 1907.

bottom: Mildred Ohrbach, my grandmother, holding my mother, 1911

Gloria Ohrbach, my mother at 18.

William Rechler, my father at 16.

top left: Gloria Ohrbach in the Packard in 1931.

top right: My parents in 1931, not yet married.

bottom: My mother and father on their honeymoon in Bermuda in January, 1933.

top: Aunt Shirley at 10 with my mom and dad and Grandpa Ohrbach at Manhattan Beach in 1932.

bottom left: Mom and me.

bottom right: Up at bat at a year and a half.

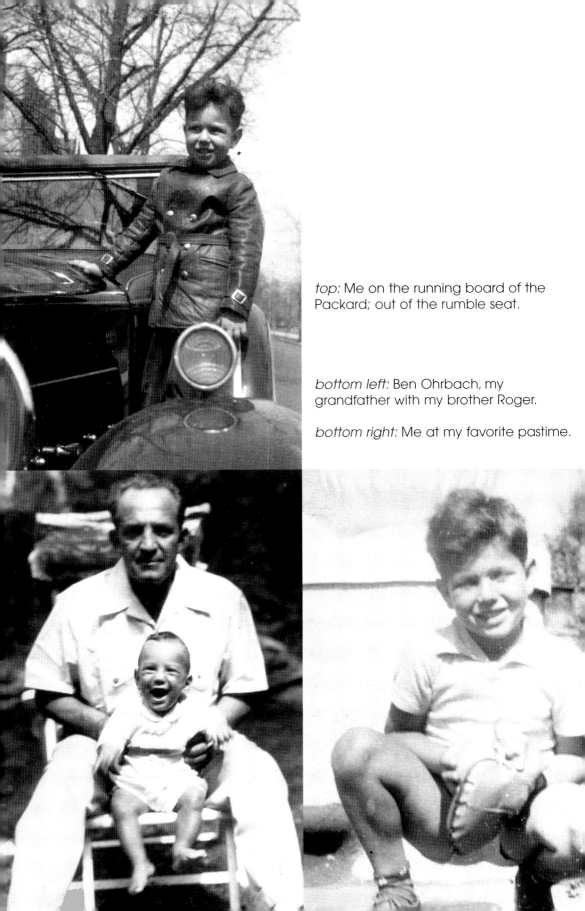

top: Me on the running board of the Packard; out of the rumble seat.

bottom left: Ben Ohrbach, my grandfather with my brother Roger.

bottom right: Me at my favorite pastime.

left: Dad and me in my soldier uniform

bottom left: Cousin Ann and my usual happy self.

bottom right: A friend, me and Sandy Koufax.

top: L-R Center 1st row: Me, Great Grandpa and Grandma Rosette
2nd row: Grandma's sister with her son, little Grandma Ohrbach;
3rd row: My grandparents, Mildred and Ben Ohrbach, Cousin Alvin;
Top row: Aunt Shirley's first husband, Aunt Shirley, Mom and Dad.

below: My bar mitzvah in the Bronx apartment. L-R Standing: My Dad Will Rechler, Aunt
Beverly, me, cousin Ann Wexler, Aunt Ruth Wexler; Seated: Grandma Yetta Rechler,
cousin Gary Wexler, Grandpa Michael Rechler, my mom, Gloria Rechler and brother
Roger Rechler.

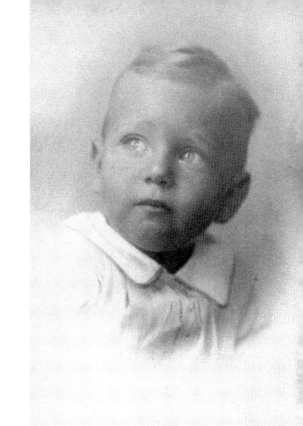

top right: Portrait of me at one year.

bottom left: My bar mitzvah portrait, age 13.

bottom right: High school graduate.

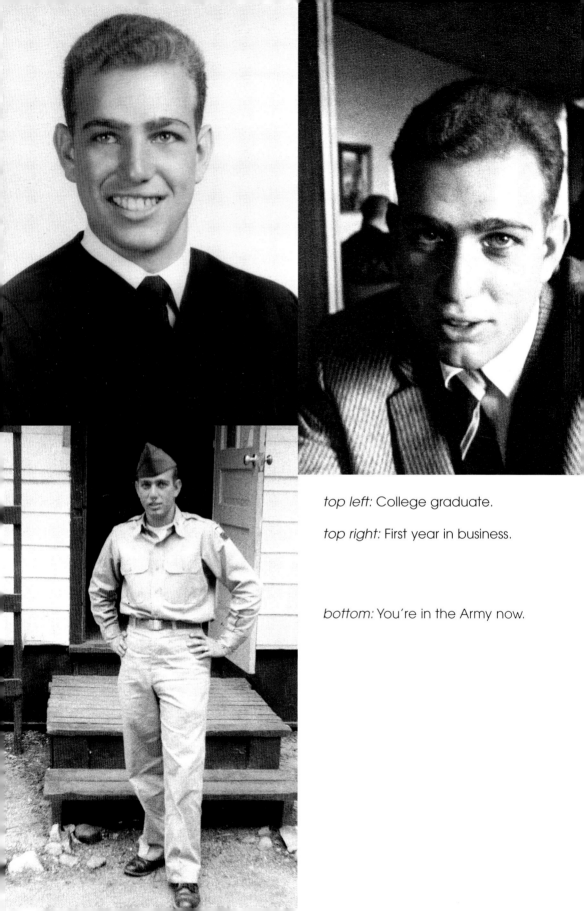

top left: College graduate.

top right: First year in business.

bottom: You're in the Army now.

top: Bruce Tucker, me and Jerry Tucker, college days.

middle: Stanley Rabinowitz and me, fraternity weekend in Key West, spring 1956.

bottom: Phi Sigma Delta basketball team. Standing left (white t-shirt) Larry Orenstein; standing right (white t-shirt) Jack Wohl; below: me center with Bruce

top: Judy when she was a little girl.

bottom left: Judy on the night she met my parents.

bottom right: College fraternity formal, Judy and me.

top: Judy meets my family in Florida in 1956; it was love at first sight (for my mom).

bottom: Judy, a few months after we were married—with that glow.

top right: Judy and me on our wedding, September 28, 1957.

top left: Wedding picture with Brother Roger, Dad, Mom and me; I think I look a little nervous.

bottom: Ushers L-R: Howard Strauss, Gary Wexler, Bruce Tucker, Roger Rechler, Stan Rabinowitz, Stan Wadler, Mickey Brager and Marvin Epstein.

top left: Mom and Dad in a rickshaw on a trip they took around the world. It's hard to believe that three years later she would pass away so suddenly.

bottom left: First prize winner: Gloria Rechler! My mother at the Mad Hatter Ball on the *SS Constitution;* she had made this hat herself.

bottom right: My mother on the steps of the *SS Constitution.*

top: My mother with Mitchell, New Year's Eve 1962.

center: Judy's father Reuben Gottlieb, Wife Lillian, Aunt Rose Katzen, Glenn and Aunt Yetty Tauber at Glenn's Bat Mitzvah.

bottom: Me with Mitchell and Glenn.

top: My father with Mark, Gregg, Scott and Todd.

bottom: The day after Mitchell's bar mitzvah, December 10, 1962. L-R: Mitchell, Mark, Grandpa Ohrbach and Glenn.

top: My father with Glenn at Mitchell's bar mitzvah.

bottom: Mitchell's bar mitzvah 1972. Our family with Joan and my dad.

top: Mark's bar mitzvah at our house in Dix Hills, 1979, L-R: Mitch, me, Mark, Judy, Glenn and Sir Winston.

bottom: Glenn's bar mitzvah 1975. Our family with Judy's dad, Ruben, and wife, Lillian, and my dad with wife Joan.

top: First New Year's Eve after marriage,
December 31, 1958 L-R: Peggy, Iris and Judy.

bottom: Mark's bar mitzvah, May 1979. L-R:
Roz Goldman, Peggy Tucker, Sandy Berlin, Iris
Rabinowitz and Judy.

REUNION TOUR

CONTENTS

Preface

While waiting for the family to come to Terrapin and celebrate Hanukkah, I started to think how my life was changing as I approached my 70th birthday. I had just left the public real estate business and would be back totally in the private sector by the first of the year. I would have more time available to me than I ever had. We were about to leave for our Boca Raton home the next day and I intended to be away for four to six weeks for the first time in my life and was wondering what I would do to fill the time.

Our family, Debbie, Mitchell, Mark, Jackie and all the grandchildren arrived and we exchanged gifts. Debbie had given Judy, Mitchell and Willi a full day of cooking with one of the top chefs in New York, David Bouley. I was a little envious of the great gift, then Debbie and Mitchell handed me an envelope. I opened the envelope and to my surprise I was given a Hanukkah and birthday gift of a biographer who would help put my memoirs together, including editing and publishing. Even though, in my mind, I had thought I might write my memoirs someday, I didn't think I was prepared to tackle that now. Debbie told me about how it worked. She had read a friend's father's book, thought it was terrific, and wanted me to meet Kate Winters who had made it possible.

I have had a great life and was in a good place in life to reflect on it. I was brought up by parents who were part of the "greatest generation." Besides my personal memoirs, I wanted to portray the dynamic era and culture I had grown up in and thrived in, as I had seen it.

I wrote this book as a series of experiences and events, as I lived them and recalled them. I painted with broad strokes. It would be impossible to condense almost 50 years of marriage and 40 years of real estate experiences in depth. I was not able to include all of my friends and family and times shared in a single volume. I chose to write my memoirs as loosely threaded highlights of events, as they wove their way into what made up my life. I spun it as a story that I would tell to a friend or grandchild while including life's lessons and "Donaldisms." I approached writing this book as I would have any new undertaking: I learn as I go. I began dictating to a tape recorder; as I reached the half way point I went to pen and paper. As those who know me, I really prefer to do things myself.

I wrote my story to share the times, the friendships, the sadness, and the laughs with family, friends and most of all my grandchildren. I was determined to do it in a book of reasonable length.

In my 70 years, I have lived through a dozen presidents, four wars, several police actions and many historic events. I have seen, read and listened to giants of politics, arts, science and sports. In that period, we have gone, as a country, from innocence to indulgence and from patriotism to skepticism. As I started reminiscing about my life from childhood to college to the army . . . and finally to my business life, and revisiting with family and friends, past and present, I felt that I was on a tour of my life and thus, this is how I came up with the title, *Reunion Tour*.

I hope this candid view helps you reminisce on your own life and reflect on lessons learned as it has for me.

CHAPTER 1

Family Heritage

"Luck is the residue of good decisions and hard work," my father would always say when someone commented on his luck. From the time I was young, I remember stories about my father being lucky. In 1943, during World War II, when I was nine years old and he was thirty three, he was drafted into the Army. They had recently raised the draft age to thirty three because they needed more men. Even though Dad had two children, he was drafted. After saying goodbye to his family, he left for the Army.

At the physical, they examined him, listened to his heart, and everything seemed to be fine. Then they sent him into this room and told him to sit there, that they would come and get him. He sat there about five minutes, hearing voices from outside the room. "Get up, walk across the room, bend down, and touch your toes." He looked around, wondering who they were talking to.

A man came into his room and told him to "get in line over there." When he got to the front of the line, his card was stamped 4F/Rejected.

"Why am I rejected?" Dad asked.

"You're hard of hearing." They were talking to him the whole time in that little room and he heard them, but he didn't think they were talking to *him*.

From this experience I learned that luck plays a significant role in life and often pops up when you least expect it, or perhaps *because* you expect it. My father's good luck followed me through my life; it became my own good luck and continues to follow me beyond this Reunion Tour . . . or was it the residue of good decisions and hard work?

*   *   *

Up to the age of thirteen, I had seven grandparents, including great-grandparents on my mother's side, which is pretty special, but I didn't fully appreciate the value of it then. As a boy, I was forever visiting grandparents, and always felt that I wasn't allowed enough time to play ball. Now I realize how much my grandparents enriched my childhood.

One of my great-great-grandfathers on my mother's side was from the Nortes of Germany. I don't remember them, but I heard stories about why they left Germany.

The Nortes were a family of bakers in Germany. As the story goes, there were two brothers whose sister was raped by two monks. The brothers beat up the monks, and one of the monks died as a result. In haste, the Norte brothers departed for the United States in the 1850s. After settling in America, one of the brothers met and married a woman of Irish-German descent. They had nine children, and settled in Pennsylvania where they ran a bakery together, and he also farmed.

When the Civil War broke out, my great-great-grandfather wasn't an American citizen, but he became a citizen when he was conscripted to go into the Civil War. He was wounded at Gettysburg and received an honorable discharge.

Shortly after the Civil War, the Norte family headed east from Pennsylvania to Brooklyn in a covered wagon—an unusual decision for the time. They felt a bakery would be successful in Brooklyn because it was a heavily populated urban setting. On the way from Pennsylvania to Brooklyn, my great-grandmother's older sister was kidnapped by Indians. This was a very common occurrence in those days, especially with girls who had long, blond hair like my great-aunt. She was never found. I recently read that the Indians would take them as wives.

My great-grandmother, the Nortes' second oldest daughter, married a German Jew named Rosette. (I believe he was born in this country or came over when he was very young). My great-grandfather Rosette worked as a cigar roller possibly in a factory, but more likely in somebody's house. Like her parents, they also had nine children. My maternal grandmother, Mildred Rosette, was the second oldest child and the oldest girl.

The elder Rosettes lived on Avenue P, right off Ocean Parkway and I used to visit them very often as a child. Since my great-grandmother came from a family of bakers she'd regularly bake 12 or 15 different things at one time, including a variety of breads. The aroma of freshly baked bread and the spicy smells of ginger, cinnamon, and cloves filled her home, just like a bakery. When she knew young children were coming she made gingerbread men and gingerbread houses and different cakes and cookies. Her own children visited on holidays or weekends. On holidays, when all the kids would bring their families, my great-grandmother set up a table that ran from the kitchen, through the dining room, through the living room, and all the way to the front porch, seating 40-some-odd people. Her children brought the food, but she did all the baking.

A big wooden table—about eight or ten feet long, which to me looked enormous—sat in the middle of my great-grandmother's kitchen. Probably it was a baking table from that time. There were no recipes or measuring cups. She poured a big pile of flour on the table, then grabbed handfuls and threw it into bowls. Baking several different things at one time, she didn't measure anything. When my mother wanted a recipe, Great-grandmother was not capable of giving it to her.

One day, when I was five or six, my mother brought me over and sat me on the corner of the big wooden baking table. I watched with interest as each time Great-grandma grabbed something to throw into the bowl, my mother grabbed her hand and made her put the ingredient in a measuring cup first. That's the way we got the old family recipes the Nortes had brought from Germany. My mother was particularly interested in three different cakes and preserved those recipes, two of which we still have in the family. One is my great-grandmother's famous peach cake and the other is her orange cake. During the holidays, my wife, Judy, still makes the peach cake today (recipe included in the last chapter of this book).

While all this baking went on in the kitchen, my great-grandfather, a thin man with white hair and a white mustache, usually sat at the end of the couch beside a big old-fashioned floor radio in the living room. His ear to the radio, he listened intently to the Brooklyn Dodger baseball games, occasionally smoking a cigar. I started listening to the ballgames on the radio with my great-grandfather.

Both of my Rosette great-grandparents lived until age 86. When my great-grandmother got sick, they knew she was going to die, and my great-grandfather moved out of their bedroom to sleep downstairs. She died in her sleep early one morning. It was discovered that Great-grandpa had already died about an hour before.

My great-grandmother had long promised my mother, as the oldest grandchild of the Rosette family, the Civil War discharge papers that had been issued to her great-grandfather. When my mother and my grandmother Mildred arrived at the house on the morning after both of my great-grandparents died on the same day, those papers were already gone. Those papers were very important to my mother as family heritage because they showed that her great-grandfather was a Civil War veteran who received an honorable discharge.

Some time later, another family treasure disappeared. My mother had the Rosette family Bible with a mother-of-pearl cover that was sent to her as a gift from an uncle in Israel. She always kept it by her bedside and even had it with her at the hospital. After she died, the family Bible was missing from her hospital room.

The Rossettes' nine children made for a large extended family and I had several great-uncles and aunts. My grandmother, Mildred, had an older brother, Willie who was quite a character. He used to race in the Albany-to-New York motor boat race and won it a couple of times. For his vocation during Prohibition he was a rum runner who ran up and down the Hudson River, bringing rum from Canada. Having the fastest boat, Uncle Willie could outrun anyone.

Another brother, Arthur, became a very successful sales manager at a large corporate firm. An interesting thing about that family is that all nine children took up the religion of whomever they married. My grandmother became Jewish, where Arthur became a Jesuit. Arthur's son later became the head Jesuit priest.

My grandmother, whose maiden name was Mildred Rosette, was an attractive woman with red hair and blue eyes. She married Benjamin Ohrbach. As he was from a Russian-Jewish family, she became Jewish when she married him, but in a very reformed way. Education was very important to her as a young woman. Before she got married, she lived in Harlem and was a hat designer at a time when every lady wore a hat. When she first married my grandfather, they lived in a fashionable apartment in Harlem.

In 1910, the Ohrbachs eventually bought a house—one of the new two-family homes being built at the time on Ocean Parkway. The address was 1676 Ocean Parkway. It was only about two-and-a-half blocks from my grandmother's mother. Her sister, Ruth, lived around the corner; another sister, Lottie, lived four blocks away; and her brother, Leon, who was an accountant lived about eight blocks away. It was quite a community of family.

People didn't move around like they do today. Neighborhoods were very family-oriented. In those days, when families moved to a house, they stayed there. My mother was born in 1912 in the Ocean Parkway house that my grandparents had bought the year before. My grandfather lived there until he was 86 years old, so he lived in the house for over 65 years.

The Ohrbachs had two daughters—Gloria, my mom, and her sister, Shirley, who was ten years younger. Those girls became my grandmother's whole life. She designed and made all their clothes, and when I was a child, she even made all of my clothes. I remember the whirring sound of her sewing machine on the front porch. A talented seamstress, my grandmother could create anything she saw in any fashion magazine and loved to use the most elegant fabrics.

My aunt Shirley, her youngest daughter, was on the wild side. No doubt the times contributed to her behavior. Shirley was seventeen in 1941 and had a good time dating servicemen. She married a young man named Herb who went into the service, not an uncommon thing to do during World War II. However, even after her husband went into the service, wild Shirley was still dating servicemen. This upset my grandmother, and she and Shirley used to get into terrible fights over it.

Shirley's problems escalated over the years. The young serviceman was the first of six marriages. Every one of her next five husbands was a bookie and she ended up printing a tout sheet for the races. The family has no communication with Shirley anymore.

I later learned that my mother had polio as a child, but I only learned this after her death. In those days, most people who got polio died, or at the very least wore braces and were handicapped for the rest of their lives. Grandma Ohrbach researched until she found a Park Avenue doctor who could save her child. You would never have known that my mother had polio because she recovered and grew into quite the athlete. After my mother's death, my grandmother told me about this Park Avenue doctor who had assured her that my mother would recover from polio, but warned her that something would eventually go wrong and Gloria would not have a long life. My mother passed away in 1961 when she was only 49 years old. She is interred in Beth David Cemetery, Elmont, Long Island. When my mother died, my grandmother hysterically told me this story, and kept repeating, "He told me. He told me."

Grandma Ohrbach died of pneumonia at the age of 72, just two years after my mother. The doctor said she simply had no will to live after she lost her beloved Gloria. My mother had been everything to her.

The Ohrbachs (my great-grandparents) had come over from Russia, but Grandpa Ben was born in Brooklyn. My great-grandmother Ohrbach, my grandfather's mother, was still alive in my childhood, so I had two great-grandmothers in my life. A very tiny, white-haired woman with stooped shoulders, my great-grandmother Ohrbach couldn't have been over four-feet-two. I used to call her "Little Grandma."

Little Grandma's husband had died earlier, so I never knew him, but in keeping with Jewish tradition, I was named after him. My great-grandfather was Daniel and I was named Donald, a more modern Irish name. My mother had a little bit of Irish in her and she was a very modern woman, quite willing to push tradition to its limit. Both names started with a "D" and that's all that counted.

Little Grandma owned and lived in an apartment house on Montgomery Avenue, about two blocks from Ebbets Field where the Brooklyn Dodgers played. Every time I walked into her apartment house—never mind her apartment—I always noticed that the entire building smelled like chicken soup.

It was about a fifteen or twenty minute ride from Grandma Ohrbach's house to Little Grandma's house, and my mother would usually pick up my grandmother and take her to visit my great-grandmother. Of course, I often had chicken soup at my little grandma's, but *wherever* I went, they all tried to feed me. In Jewish homes, that's what grandmothers do.

Growing up, we celebrated Thanksgiving and Christmas at Grandma Ohrbach's house. She was an excellent cook and it was a trait she passed on to my mother. Grandma Ohrbach always had a Christmas tree, and to me, this was novel. My grandfather's ancestry was Jewish and my grandmother's was Christian. Although she'd converted when they married, they exchanged Christmas presents simply because it was a tradition.

It was not a tradition in our house. When I was a little boy, I remember my parents having an argument when my mother brought home a little Christmas tree. My father said, "absolutely not," and nixed the whole idea, but I enjoyed the tree at my grandmother's house.

It was Grandpa Ben's cousin who started Ohrbach's Department Store, a very large chain in New York, but my grandfather was interesting in his own right. He started the first taxi fleet in New York City. He had the first New York Taxi Medallion. Not only did he start the taxi service, but Benjamin Ohrbach was actually one of the drivers. In fact, he used to drive Jimmy Walker around all night. Jimmy Walker was the New York mayor at that time and a real *bon vivant* personality. A celebrity in his day, Walker carried on with showgirls at nightclubs and speak-easies.

We all used to call Grandpa Ben *Yankee Doodle Dandy* because he was just the perfect gentleman. He would always jump out of the car and run around to open the passenger's side door to let a lady out. I remember when he was 84, he still managed to get to the other side of the car to open the door for Judy even though every step looked painful.

My grandfather eventually owned a garage with a fleet of cabs. At that time, certain industries were targeted and being taken over by gangsters. Ben had trouble because his drivers kept getting beat up. The mob wanted him to pay them protection. I think between that and the Depression, he found it too difficult to run the garage with the fleet of cabs and to manage all the drivers, so he kept his own cab and did his own driving from that point on. He always made a very good living. I say that because many years later, Grandpa Ohrbach would lend Judy and me the money to buy our first home in Jericho, Long Island ($6,000, a lot of money at that time).

The paternal side of the family was originally from Yasse, Romania. Goldstein was the family name of my father's mother. My grandmother, Yetta Goldstein, married Michael Rechler who was my paternal grandfather. His parents were the ones who lived in Yasse. All I know about the Rechler family in Romania was that they were the wine merchants of Yasse, so they were fairly well-to-do and respected.

There are two stories about how my grandfather, Michael Rechler, met my grandmother, Yetta Goldstein. The one I heard from my parents was that he had to leave Romania and the family business in order to marry my grandmother. Apparently, if you had money when you served in the Romanian Army then you could pay the sergeants or other soldiers to do all your work for you. After his discharge, Michael Rechler went to live with his darling, Yetta Goldstein, who lived on the wrong side of the tracks. In those days, families chose who they wanted their children to marry.

My cousin told me another story that she heard from my grandmother. She said that my grandfather had to leave Romania because his family didn't want him to get married, even though he was twenty-seven at the time, because his two older sisters weren't married yet. My grandparents' real reason for leaving Romania may have been a combination of the two stories, as both sound plausible. In any case, they left their home and their families to be together.

In 1905, my grandfather, Michael Rechler, came to America not only with his wife, Yetta, but also with his brother, Ben. They settled in Brooklyn, New York—not exactly wine country. The Rechler brothers started out by selling piece goods from a pushcart on the Lower East Side. They finally worked their way up from pushcarts to piece goods (fabrics) stores in Brooklyn, and they both made a very good living.

Their first son, Daniel, died when he was about fourteen years old. The way I heard the story, Daniel accidentally shot another boy's eye out with a BB gun and he took such a beating from my grandfather that three or four years later he passed away. No one was sure the beating was the cause of his death, but they say Daniel was never the same. The Rechlers also had twins that died of influenza when they were barely a year old.

My father, William Rechler, the second son of Michael and Yetta, was born in Brooklyn on May 6, 1911. He had a younger sister, Ruth, and a much younger brother, Morton. That was the Rechler family in Brooklyn.

My grandfather vowed never to go back to Romania until he was successful. It was over 16 years before he and his brother, Ben, took a ship to Yasse, Romania. He left my grandmother for six months with all their children, including my father—then only a sixteen-year-old—to watch two retail stores. It was a different world at that time. The two Rechler brothers went back to Romania around 1927, showing everyone how successful they'd become in America, where the streets were lined with gold.

Many years later, in 1983, I heard a story about the trip the two brothers took to Romania. I had been in real estate for about three years. I'd learned that a large linen chain called Frankel's, with stores in Garden City, were looking for a warehouse, so I made an appointment to see the owner, Bob Frankel. When I arrived at the store and introduced myself at the reception desk, I heard this heavily accented voice bellow out, "Are you Michael Rechler's son? Come in."

"No, I'm his grandson," I told him. My grandfather had passed away by then.

"Oh, I remember your grandfather," the older man, who was Bob Frankel's father, said with a far away look in his eyes. "I came to this country because of Michael Rechler. He and his brother were in Yasse. They wore white suits with straw hats and vests and looked so splendid with their canes. They said they were millionaires and told us all how great America was. Half the town followed them back to America." That kind of scenario happened often in Europe then.

Whenever I saw my grandfather, he was always dressed in his Sunday best. An elegant lady, my grandmother was always well-coiffed and well-dressed too. My grandparents spoke Yiddish a lot, especially when they didn't want the children to know what they were talking about. My grandfather taught me some Yiddish, such as *cocki muon* which means "go shit in your hat." Not that I know much Yiddish today, but it's a phrase I remember from my grandfather.

I understood my grandfather could be a difficult man, but he was never difficult with me. My grandmother was the mediator who often had to smooth things over with her children's father. She encouraged her children, but expected a lot from them. Yetta Goldstein Rechler was very much the matriarch of the family, and everyone looked up to and relied on Grandma Yetta. My father's sister, Ruth, and her husband, Jack, (and my cousin Ann) lived either with or next door to the elder Rechlers for most of their lives. The relationship between Yetta and Ruth was inseparable, as Ann's later became with her mother.

The Rechlers raised their family in Brooklyn, but later moved to the Bronx on Pelham Parkway, where my parents and I had dinner every Sunday night. I was the oldest grandchild in this family too. Aunt Ruth was always there with her husband, Uncle Jack. There would be more kids later, but it was just me and their daughter, Ann, at first. Uncle Mort, my father's youngest brother by 11 years, eventually married Aunt Beverly.

After our pot roast or chicken served family style, every dinner ended the same way: in a raucous debate that usually turned into an argument. Whether it was about politics, intellectual arguments or sociology, the heated discussion was usually led by Uncle Mort and Aunt Ruth who were both bright, well-read people with strong beliefs. My father and mother participated, but they weren't as combative. My grandmother was involved in the conversation, but after dinner my grandfather would always get up and go into his room. I'd follow him. He'd sit in his chair and light a cigar, while I sat quietly next to him and listened to his war stories. I'd heard them all before, but they were comforting. I'm sure those fond memories have encouraged me to share my stories with grandchildren and family.

My grandfather Rechler eventually gave up his stores in Brooklyn and opened one very large store which he owned for many, many years on Third Avenue and 149th Street in the Bronx where the "El" (transit system) was at that time. Reckler's (spelled with a "k") was the largest small department store of its kind, selling piece goods. Across the street there was an upstart that used to copy him but also added women's dresses. That store was Alexander's, which became a huge discount department store run by Farkas.

By then, my father was in the business, running the lower floor at Recklers, which stocked gifts, household items, lamps and some furniture. I sometimes accompanied him to the store, and as a kid, I worked there at Christmas time. I clearly remember my grandfather at the store sitting upstairs in his usual spot at the store every day, right by the stairs where he could see the front door, the cash register and everyone going up and down.

"If you don't like your work, walk down the street and smell the butcher shop." —Michael Rechler in a letter to my father.

As they became older, my Rechler grandparents vacationed in Florida in the winters, and they'd stay at the Lido or the Shore Club. When I attended college in Miami, I used to see them in Florida. Feisty, elegant Grandma Rechler loved to go to the racetrack and bet on the horses. A vivid memory is of the time my roommate, Bruce Tucker, and I took Grandma Yetta to the races.

She was all dressed up as usual, even wearing a hat. Excitedly, she jumped right into the action and bet a daily double (where you have to pick the winners of two races). She purposefully chose a horse with ninety-nine to one odds in the second race, and one with a forty to one shot in the first, so the payoff would be something like four to five thousand dollars for her two dollar bet.

Amazingly, her horse won the first race! In the second race her long-shot horse was leading all the way, but right near the finish line something happened; it bumped another horse and was disqualified for a foul.

Grandma Yetta refused to leave the racetrack until she got her money. Bruce and I simply could not explain this to her. We finally took her to the betting window and let them explain it to her, and she was only convinced after we took her to two or three windows and each time they gave her the same explanation.

Grandma Yetta passed away in 1962, before my grandfather, at about the age of 72. Her husband (my grandfather) Michael Rechler, later died of pneumonia when he was 86.

CHAPTER 2

Will and Gloria

My parents grew up in the 1920s, in the flapper days, the Prohibition days. My mother wore the most fashionable clothes of the time, just like they wore in the movies. Since my grandmother used to make all her clothes, copying what she saw in magazines and on the silver screen, my mother dressed well even before she could afford it.

Born in Brooklyn on December 2, 1911, my mother, Gloria Ohrbach, grew up in Brooklyn, New York at the house on Ocean Parkway. She was an only child for several years and was very much my grandfather's son as well as being my grandmother's daughter.

Grandpa Ben was a terrific athlete. A strong swimmer, he won the long-distance races at Manhattan Beach and medals in the breast stroke swimming races for many years. He also reigned as the handball champion of Manhattan Beach. My mother followed his footsteps in both sports. She won the women's breast stroke race and the women's handball championships in Manhattan Beach. Several thousand people used to enter those contests so it was a big deal, especially the handball.

My mother was talented in the arts, which she probably got from my grandmother. She was brought up during the

Depression and my grandmother wanted her to be educated in everything, so she taught her etiquette, and gave her piano lessons and ballet. After she finished high school, she went to New York University.

My parents met at NYU. At the time, my mother was dating the captain of the football team. The way I heard the story, because my father wanted to meet her, he joined the football team, and even though just a substitute who sat on the bench, he eventually won her over. She was young when they met, barely eighteen. Although he was only six months older, my father always looked very mature (his picture at age sixteen in the photo pages makes him look mid-twenties). That's how, when he was only sixteen, he got away with running the two stores when my grandfather and his brother, Ben, went back to Romania.

My father was probably the only one who got left back in kindergarten. Apparently, when he was three years old my grandmother put him in kindergarten to get him out of the house. There was no day care or nursery school then, so she used it as a baby sitting service. He stayed in kindergarten for three straight years!

Another funny story I was told about him happened when he was a little boy. As he walked the several blocks home from school one day, he was talking and a fly flew in his mouth. He didn't know what to do, so he shut his mouth and ran the rest of the way home with the fly buzzing around in it. When he got home, he pointed urgently at his closed mouth. His mother made him open it and out came the fly, still buzzing.

Dad was athletic, but not the kind of athlete my mother was. He ran track and won medals, and although he made the NYU football team, he never got to play in any of the games.

He majored in business at NYU, a subject he'd focused on from a young age since he already worked in the family retail store, and did well. In the Depression years, business was conservative, guided to money, cost and savings. In those times every penny counted and risks were not taken.

At fifteen or sixteen, my father was already driving. He later told me about the day he accidentally hit a horse and cart, killing the horse. He just didn't see the black horse in time.

Just recently I heard the rest of the story from my uncle Mort, the part my father didn't tell me. When he arrived home on the night he'd killed the horse, the peddler who owned the horse, showed up at the door with a policeman. My grandfather opened the door, and my father and uncle were right behind him.

The policeman announced, "Your son hit this man's horse and killed it."

My grandfather asked, "Villie, did you do dat?"

"Yes, Pop," my father said.

The policeman said, "We won't press charges, but you have to pay the man $50 for the horse."

"Vat is dis? A race horse?!" my grandfather said. He reached in his wallet, took out the $50 and handed it to the man. As he brought his hand back, my grandfather hit my father across the side of his face with the back of it so hard that he knocked him down.

My parents both loved horses and would later enjoy riding together. As you can see by the pictures in the photo pages of the two of them dressed in riding habits, they rode together even before they were married.

When my father started taking out my mother, his parents weren't happy because in their minds she wasn't really

Jewish. She had a Christian-born mother. Yet she had been confirmed in the mid-1920s, which was almost unheard of in those years for a Jewish girl. It was a reformed ceremony, partly due to her mother's heritage and partly because she was brought up in the first reformed synagogue in Brooklyn. That is how she brought up Roger and me, and we continue that tradition today.

My grandfather Rechler was Orthodox. He always went to temple on the holidays and on many Saturdays. Unfortunately, every once in a while he dragged me there with him. It was like culture shock to me. The women sat by themselves in a balcony, separated from the men. The men were all chanting from what seemed like different pages and it sounded chaotic. They stood most of the time, so what I saw was a lot of behinds.

My parents started dating in 1930. They married on January 1, 1933—New Year's Day—and honeymooned in Bermuda. When they returned they moved into their first Brooklyn apartment, where I was born about twenty months later.

My mother never worked. She went to school and then she was married. In those days, married women stayed home and took care of their families. A very, very good cook, my mother enjoyed following and collecting recipes and cookbooks. The Brooklyn apartment, where I lived for my first six years, was near her mother (my grandmother), her aunts, and her grandparents. That was good because family played a big part in her life.

My father was a retailer who worked every night until ten o'clock, so he never ate dinners with us when I was a young child. He worked six days a week, especially during the

Depression, and that didn't change much during the war years. In those days, it was hard to get merchandise for the stores. The government only allowed necessary items to be made during the war. Thanks to my father's ingenuity, Reckler's was able to get product and survive the times when many stores didn't.

As an example, in order to get porcelain lamps (which were not considered necessary items during the war) they had to be fired with toilets (which were). It worked like this: toilets are porcelain. Porcelain goes through a kiln that uses fuel. At the time, the government didn't want fuel used unless it was a necessity. He would ask the manufacturers to put the lamp in the kiln with the toilet, so they both could be fired at the same time. He had other products fired up the same way, so he prospered during that period.

An amazing thing about my mother was that she was remarkably intuitive. People used to call her a witch, but in a good way. Gloria Rechler could predict things, not so much as a visionary, but she just knew what was going to happen around her. Being intuitive wasn't something she studied or even nurtured, it was a natural talent that she was quite matter-of-fact.

Another memory of my mother is when my grandmother used to call her every single day and probably stayed on the phone anywhere from an hour to two. My mother would chat for a few minutes, then go about her business and just keep the phone on the table. Every once in a while she went back to the phone and said, "Yes, Ma. Okay, Ma."

I would say that my mother's love for art and my father's collecting of art greatly influenced my life. All of the Rechlers were collectors. Aunt Ruth and Uncle Jack were tremendous collectors. When my father could finally afford it, he collected paintings. Before then, wherever my parents would travel, they would collect something, much like Judy and I have done. My father collected ceramics. My mother loved the Art Deco period—the 1920s and 1930s.

My father was eclectic and his art collection was really varied. He had a number of paintings from the Ashcan School, a legitimate period of art, which consisted of New York artists who painted during the Depression. He knew some of the artists, but the paintings were not extraordinarily valuable. After my mother died, my father visited Mexico and made friends with some artists there. He started to collect Mexican artists, such as Orozco and Siquiros. He also collected French Impressionists like Mary Cassatt and Vlaminck.

My mother was in her early forties when they moved to their house in Kings Point, Long Island. It was then, in the 1950s and 60s, she started taking art lessons, and became very good at painting and sculpting. She studied with some of the famous artists in the area; some of them were the Pop artists who started with the first huge canvases, very abstract and different (Pop Art). She liked abstract art, but Dad couldn't relate to it then.

My mother made friends with all the artists and loved to be around arty people. One of her teachers was Larry Rivers, a famous pop artist in the same school as Warhol and Rauchenberg. One time, he gave four paintings to my mother in exchange for some money to live on. When she brought the canvases home, my father looked at the large, abstract paintings, and said, "What are you taking these for? Where are we going to put them? When he can pay you back, you'll take the money. Until then, let him keep his paintings." He made her return them, which she did.

Larry Rivers was unknown then, but later those very canvases became worth an awful lot of money. The funny twist to this story is that after my mother passed away my father got more and more into art and even opened up his own hobby art gallery. He went on excursions to see the newest art, and became a collector of Pop Art (even though his new wife hated it). He ended up collecting what my mother had brought home and what he had rejected years before.

I was still in high school when my parents designed and decorated the Beechhurst, Long Island house themselves and moved us to the country. Just two blocks up from the Long Island Sound, our new home was very pretty and they loved it. But it was a completely different world and a difficult adjustment for me.

At that point, my father had left the retail business to start a manufacturing business with my uncle Mort. They bought an old shipyard building in Whitestone, Queens, and launched a manufacturing business that lasted for about nine or ten years. My parents decided to buy the Long Island house because it wasn't too far away from Whitestone and College Point where the plants were.

It all started when Uncle Mort left the Air Force, where he was an engineer. While in the service he'd fallen in love with aluminum, a new material at the time which the Air Force used to make planes. He and my father thought aluminum had great potential for manufacturing, so they looked for just the right products to make out of aluminum. At first, they made trays and glass coasters or caddies, as they were called. Finally, my father suggested, "You know, what there needs to be? An aluminum folding chair."

They made the first aluminum beach chair and aluminum furniture. The company was named Rexart, and the folding beach chair was called Compacto. The chairs took off right away and Rexart immediately grew and became successful. Towards the end, as a major supplier for Sears, Rexart's annual volume was six million dollars. Sears eventually became their downfall because when the Sears buyers purchased such high volume, they could easily dictate price and overhead. But I would learn from this lesson.

"As a boy you imitate your father. As a teenager you avoid your father. In your twenties you laugh at your father. In your thirties you appreciate your father. In your forties you become your father." — Donaldism.

My father was very industrious and innovative. He was a dapper gentleman, a "Mr. Nice Guy" whom everyone liked, and a very good salesman. Happy by nature, Will Rechler loved to laugh and was always smiling. He enjoyed parties, loved to give them and loved to attend them (which I think is an inherited trait in my family). A social drinker, he never had more than a couple of drinks. There were six couples in my parents' group of close friends, and they all took turns having dinner parties at their houses in later years.

In the mid-to-later years at Rexart, my father experienced a lot of frustration and frazzled nerves, probably because he had worked so hard; it happened again at the beginning of his real estate career. He was a people person who could sell anything. In real estate, he was the one who dealt with the tenants and brokers.

Dad was the older brother by eleven years and easily got along with most people, but my uncle Mort was a very dogmatic man. As an officer in the service, he was used to getting his own way. Compromise was not one of the words in his vocabulary, so my father had trouble dealing with him. When Dad got high blood pressure and broke out in hives I overheard my mother encouraging him to do something else, to separate himself from my uncle. It probably took a couple of years for him to do that.

My father's philosophy was, "Do what I tell you, not what I do." He told me that I questioned every lesson he ever gave me in life or else I pointed out to him that *he* didn't do it that way himself. I called him Dad until I started working with him, then I called him Will like everyone else did.

CHAPTER 3

Early Years

I was born on October 10, 1934, and my brother, Roger, was born February 23, 1942. Roger and I were surprisingly close for brothers with that much difference in age. One reason was that I took him with me when I went out to play ball; he was always the little brother who tagged along.

Named Donald (middle name Joel, which I never use), I was called Donny as a child. I was a platinum blond baby with blue eyes, but later on my hair turned red and my eyes turned hazel. I have been told I was a spoiled brat and, considering the circumstances, I probably was. I had the finest carriage and the finest clothes. I spent a lot of time being catered to at my grandparents' house on Ocean Parkway.

Times were hard when I was brought up during the Depression. In my early years a doctor told my mother I was anemic. I wouldn't chew meat. During the Depression meat was very hard to get and during the war years it was rationed. My parents would get their one steak a week with their ration stamp and, instead of eating it, they put it in an orange juice squeezer, squeezed out all the juice, and fed it to little anemic Donny.

My earliest memories of childhood are traumatic as most early memories are. Amazingly, my earliest memory took place when I was twenty-two months old; that's young to have such a vivid memory. I could describe everything about the room where the trauma happened. We stayed for only one summer in Rockaway, so my age can be pinpointed exactly.

We spent that summer with Grandma Rechler, my father's mother, Yetta. My parents had gone out that night and left her to babysit. I was probably crying in the crib. My grandmother took me out of the crib, put me on top of a high dresser and shook me. That petrified me, so I screamed more and more. This memory is so clear that I could describe the room in very minute detail, as well as everything about the incident.

Another early memory occurred in 1939 when I was four years old and visiting my Ohrbach grandparents on Ocean Parkway in Brooklyn. I was definitely Grandma Ohrbach's favorite grandchild; I was the only one then. Growing up, I spent a lot of time at my grandparents' Brooklyn home. I suspect my mother and father brought me there to get rid of me for a weekend or a Sunday when he was off work. I stayed on their front porch (they called it a front porch, but really it was a closed-in room), right on busy Ocean Parkway. Grandma gave me a writing pad, and had me count the different color cars or number of horses that passed. As well as being a three-lane road, which was unusual for those days, Ocean Parkway had a bridle path and a bike path, but there were more cars than horses.

Aunt Shirley was around sixteen or seventeen, and she still lived there. She always stayed out late, so she slept during the day on a couch on the porch. One day, my grandmother was inside cooking. Bored with counting cars and carriages, I was playing absently with a little toy car that ran on a round wheel on a table. While my aunt Shirley slept, I ran the car over her head. It yanked out a big chunk of her hair. I was frightened and started to scream. Aunt Shirley jumped up, but instead of crying out, she began laughing hysterically. In those days, women used to add clumps of hair called rats into their own hair to make it fuller. So it wasn't really her hair that I pulled out. It was her rat.

Every time I stayed with Grandma Ohrbach, she made her special coffee-milk for me–just a little bit of coffee, heavy cream, and sugar. She always made toast for me on this very old Westinghouse toaster that toasted one side of the bread at a time. A saver, both by nature and because she'd come through the Depression, she saved buttons, strings and everything. In the late 1950s her toaster broke, so she sent it back to Westinghouse to get it repaired. At this point, the toaster was 35 years old. She got a letter from the chairman of Westinghouse, telling her how pleased the company was that she still had it, and that it would take a little time to repair it but they wouldn't charge anything for the repair. They sent her another toaster to use in between and asked her if they could take a picture and use her name to tell about the Westinghouse toaster. Sure enough, she didn't want any other toaster—she wanted *her* toaster. So Westinghouse repaired it and sent it back to her, and she made her one-sided toast again. I remember it was the best toast you could eat.

Maybe the most traumatic, indelible memory I have of childhood happened when I was about 7 or 8. My mother had punished me. I don't know why, but do know I was furious, and she wasn't going to let me out of the house for a week.

Feeling her punishment was unjust, I ran into my room and slammed the door.

"I hope she dies just like this bulb goes out," I vowed in a fit of anger. As I turned on my light switch at that same moment, the electric bulb in the ceiling of my room exploded and sparks shot out. I dropped to my knees and prayed. I was hysterical for about 15 minutes, praying: "I didn't mean it . . . . I'm sorry . . . ."

When I got up the nerve, I walked outside to see if my mother was okay. She was, and I went back into my room and thanked God. From that childhood experience onward, I have never doubted the existence of some sort of higher being. It didn't make me a religious person. I simply accepted without question that there are things we don't understand.

Almost every nice day during the summer time, my mother put me in the Packard convertible, stopped to pick up my grandmother, then drove to Manhattan Beach. In those days, a Packard convertible had a rumble seat where today's car trunk is located. My grandmother rode in the front seat with my mother and I rode in the rumble seat. I must have weighed all of 45 pounds and, about half the time on the way to Manhattan Beach, Mom hit a bump and the rumble seat closed. I would be locked inside and she wouldn't notice until we got to Manhattan Beach, then they'd let me out. It frightened me the first time, but I got used to it.

In the late thirties, Manhattan Beach was *the* happening place; beach lovers in Brooklyn went there. I went there regularly from 1937-1942. There was a beautiful bandstand with a huge concrete dance floor in front of it where people used to jitterbug during the war years. All the big bands played there: Benny Goodman, the Dorsey Brothers, Harry James . . .

my favorite was Al Donahue. My grandmother gave me a little spiral notebook to get the autographs of different musicians, and I stood at the stage door when they were done, waiting for them to come out. After collecting autographs all summer I lost the notebook and was very upset. My grandmother calmed me down. She bought another notebook, asked me the names of all the autographs I'd had, then proceeded to sign them in the new book. I was happy, but I guess I wasn't too smart!

Manhattan Beach was one of the great beaches, a wide beach with a lot of sand, probably as big as Jones Beach. The huge beach was absolutely packed on weekends and holidays, every square foot of sand occupied. A lot of people liked to sit by the ocean, but my mother liked to hear the music and watch the shows, so we sat up front by the dance floor which was all open. On weekends, the bands played during the day. I often watched my parents dance together on Sunday afternoons in the summertime.

I also once saw my mother compete in a dart contest against thousands of people. After two weeks she won in the women's division. Then the organizers wanted to hold a contest with the men's winner versus the women's. "Want to see how you do against the men?" they asked my mother. My father was there that Sunday. Sure enough, she beat the male champion. She was always very competitive. Dad knew she would win, but thought she should have declined the challenge.

One very hot day at Manhattan Beach, I remember that my father took me to the concession stand because I was thirsty. With me perched on his shoulders, he told my mother we were going off to buy a soda. First, he stood in a long line to buy a beer. When he finally got in the soda line, which was even longer than the beer line, he was tired of waiting. He stood in line for awhile, drinking his beer with me still on his shoulders. I hollered that I was thirsty, so he handed me the beer and started to walk back to the beach. I was so thirsty that I didn't give his beer back to him right away and the next thing I knew, Dad was stopped by a policeman and given a ticket. I'd almost gotten him arrested for contributing to the delinquency of a minor.

Whenever my father went to the beach he always had to struggle opening up the beach chair. It would take him 10 or 15 minutes to open it, and in the process he'd throw it on the ground a couple of times in frustration. His struggles with those beach chairs ultimately led to his successful business venture with Rexart and the aluminum beach chairs.

Every Sunday, we visited my father's family in the Bronx. Their apartment overlooked Pelham Park where I liked to sit next to my grandfather Rechler, listening to his stories. My grandmother often made a Romanian dish of chopped eggplant, onions, and olive oil, which I ate on crackers. She would scorch the eggplant before she chopped it up, and I liked the taste of that.

On other occasions we visited Little Grandma, my great-grandmother Ohrbach on my mother's side of the family. I always had to have her chicken soup the minute I walked into her house. As I mentioned, I had a lot of grandparents.

The grandmother we visited the most was Grandma Ohrbach. I stayed with her a lot and her specialty was to fatten me up. She would make me *gribenes*, rendered chicken fat with onions, and serve it on a piece of bread. At that time, it was the best thing I ever ate. Unfortunately, it's still repeating on me some 60 years later.

When we visited the nearby Brooklyn grandmothers, we often took our beautiful red dog, Ginger, a large chow that was part spitz. Ginger couldn't be let out by himself because he'd take off after anything. One time, while visiting Grandma Rosette, the dog had to go out, so I put the leash on him and we went for a walk. Ginger saw a cat half a block away and started to run. I was holding his leash and he just pulled me along. I wouldn't let go of the leash because, even at five or six years old, I knew he would run away. I held onto that leash with all the strength I could muster and the dog pulled me about two and a half blocks. Finally, my knees bloody, I had to let go of the leash. Ginger took off and we never found him again.

The interesting part is that, about six months later, his picture appeared in several papers with Lorraine Day, a famous actress. The story reported that a wild dog had been seen roaming in Prospect Park in Brooklyn. At first, they thought it was a wolf. The story made the papers because Lorraine Day went to the park and fed him every day. The dog grew tamer and would sit near her in the park. Gradually, she won his confidence enough to take him home with her. The story with his picture appeared in the newspaper. That was our Ginger, our link to *"Six Degrees of Separation"* from fame.

I spent most of my time with adults up until I was six. As I said before, my mother was always taking me somewhere. From the time I was three or four she took me to the movies. She took me to all sorts of adult movies, like Edward G. Robinson shoot-ups and gangster films of that day. Sitting through *Gone with the Wind* for three and a half hours was especially difficult. Later on when I was a little older, I would see two feature films and five serials and six cartoons. (Your mother could park you at the movies for several hours on the weekends. Child care centers weren't necessary.) I saw so many movies as a kid that going to the movies is really not my favorite thing to do today.

I loved baseball as a kid, and Uncle Charlie, my grandfather's younger brother took me to the games at Ebbets Field. He lived in Little Grandma's apartment house, a couple of blocks from Ebbets Field. After serving in World War I, Uncle Charlie got a pension because he had been gassed in the war. Basically a "good time Charlie," he sold Wrigley Chewing Gum as a sideline for a little extra money.

Since Uncle Charlie knew everyone at Ebbets, they'd give him seats right on the field, so I had an up close and personal look at baseball and the Brooklyn Dodgers from a very early age. There wasn't a bad seat in Ebbets Field anyway, but being right on the field like that was like being part of the game—very exciting to a young kid enamored with the sport. Times were different then, and the ballplayers all lived in the neighborhood in Brooklyn. When I went to the grocery store, the shop owner or Uncle Charlie would introduce the players to me; even they became almost like family.

Those were the days without computers or television, but there were lots of radios and eight different newspapers. In the 1930s, there weren't a lot of sporting events to attend. Horse racing and boxing were two of the biggest sports, along with baseball. Basketball and football were really college sports at that time. There were professional teams, but no one paid much attention to them then. In the early 1940s, still early for football, my father took me to a Giants' football game at the polo grounds. Irving, a good friend of his, had season tickets.

My first favorite player was Ward Cuff, then it was Tuffy Lehman—both running backs. I even had a Tuffy Lehman doll.

Around the beginning of 1941, just before World War II started, we moved from Brooklyn to the Bronx. We lived at 325 East 176th Street, which was about three blocks off the Concourse, the major roadway in the Bronx like Fifth Avenue is in New York City. I was about six-and-a-half years old. There we lived on the second floor of a brand-new apartment house.

A two bedroom apartment featuring a sunken living room, our first Bronx home was designed and decorated by my mother and father. My father worked at the Bronx store then, so this was convenient for him. It was of the Bauhaus School (German Modern), which was really progressive for 1941 when you realize they decorated a sunken living room with a black rug and large red Bauhaus love seats. The Bauhaus School, a period of architectural design and furniture that came after Art Deco, was extremely modern. But then again, my mother was a very modern woman. When people came to visit they were amazed because no one lived in a home decorated like ours. My brother and I still have a few pieces today. They were not very expensive in their day, but today, because of the design element, they're worth a lot of money at auctions.

Stanley Becker, a boy about my age lived right across the hall on the same floor. Stanley and I became friends and shared a love of baseball. By the time we were eleven and twelve years old, we went to the polo grounds—about two stops away on the subway—to get the players' autographs as they came out of Giants' Stadium. We always went when the Dodgers were in town.

I got very good at math through reading the baseball statistics. Each day I'd do the statistics in my head, so I knew when the players got their three hits how much their batting average went up or down. The next day I checked the newspaper to see if I was right. It became a game to me. Eventually it became easy to do all sorts of numbers in my head, a talent I used pretty much all through my life.

In the Bronx I developed my first close boyhood friendships with both Stanley Becker and Walter Hirsch. Stanley was called *Fat Stanley*, because he was chubby then and ran very slowly. Wally was called *Wally Gator* because he had two big front teeth. Everybody had a nickname in those days, including all the ballplayers.

Bubble gum cards were the big thing for kids then sold at local candy stores. I had a very large collection of baseball cards and comic books. We traded them between us. You'd flip the cards against the wall, or trade them. At a very young age I honed my negotiation skills, necessary in trading the baseball cards I had for those of my Brooklyn Dodgers' players.

There were a lot of kids in that Bronx neighborhood and we had enough friends to play stickball in the street. We would play from sewer to sewer, using a broomstick and a pink Spalding ball, which they called a high bouncer. A school-age boy never walked anywhere without a Spalding ball in his pocket. That was standard gear.

In the Bronx, we played ball all the time, but it was hard to get a proper organized game together on a real baseball field. When I was about eleven, we got enough kids together to make two teams, and after planning the event for weeks, we all took a subway the three mile distance to Van Cortland Park where there were plenty of fields. All excited, we brought our baseball gloves and equipment. The park in the Bronx was very rocky, so to get a flat place like this was great.

We'd been playing for about half an hour when out of nowhere these two large, tough-looking teenagers, much older than us, walked onto our field. We quickly grabbed our stuff. Living in the Bronx, we were wary of incidents. One of the guys pulled out a knife and the other flashed a zip gun. (Gang members made zip guns by converting the spring action of a gun used to shoot balsa wood airplanes so it could shoot a .22 shell. It wasn't a bullet, just the shell.)

The leader bully asked us for whatever change we had. We were street-wise enough to keep our change separated in different pockets, so we gave them some of it, hoping we still had enough for the subway back home. Then they looked at the gloves and demanded, "Give us your gloves."

I had two gloves, but I was only holding my first baseman's glove and had hidden the other under my shirt. One of the bullies reached out for my first baseman's glove and I held onto it as I backed away from him.

"Drop the glove on the floor now or I'll shoot you," he barked.

"No," I said firmly, knowing that while the zip gun didn't hold a bullet, it did have a BB-like shell that could seriously cut into me.

He shot me in the stomach. Right where I'd hidden my other glove.

Seizing the opportunity, I grabbed my stomach and fell to the ground. The gang toughs got frightened and ran away. I ended up with only a slight hole in the padded part of my baseball glove and none in my stomach. My glove had saved me.

Our main entertainment then was the radio, not like today where TV is second only to videos and DVD's. We listened to baseball games and serials like *The Lone Ranger* . . . and *Green Hornet* . . . and *The Shadow*. During the war years, of course, all the super-heroes like Captain Midnight were fighting the Nazis and the Japanese. All the comic books were geared to that even before the war.

I was seven years old when Pearl Harbor was bombed. It was a Sunday and I was in Brooklyn with my father, visiting Uncle Benny. Uncle Benny had been listening to a program on the radio. It was interrupted so Franklin Delano Roosevelt could address the country. Gathered around the radio in stunned silence, we learned that the Japanese had bombed Pearl Harbor. I heard Roosevelt's famous address: *"This day will live in infamy . . . ."*

I didn't feel fear; I felt anger and a sense of adventure. After all, I'd been reading the comic books and hearing the super heroes on the radio. I believed we were just going to win. When I heard the president, I only thought of victory and didn't consider the possibility that anyone on our side might get killed. That realization came to me later.

Every night at six o'clock the war news came over the radio. Just like I listened to the sports at seven o'clock, I listened to war news at six. They would announce the casualties of the day. During the war, people put up banners in their windows. A blue star meant they had a son in service. If they had three sons, they'd have three banners about 12" x 14" hanging in their window. A purple star on the banner indicated a wounded son. If the banner had a gold star, it meant they had a son who died in the war. All of a sudden, a lot of purple and gold stars began to show up in the windows. With that came my realization of the sadness of war.

Awareness of the reality of war came to me in another way too. At the age of nine or ten I was sent to my first sleep away camp—Camp Swago in Massachusetts—for eight weeks. I didn't enjoy camp because they rushed everything. You'd go to an activity, and by the time sides had been chosen, there was hardly any time left to play. If it was baseball, you played an inning, then moved on to the next activity. I hated that.

Our two counselors were veterans of the war who had been wounded in the Battle of the Bulge. They became my first close confrontation with what the war does to soldiers. One, a good-looking blond, blue-eyed young man had part of his arm missing because it had been blown off by shrapnel. His arm was scarred right behind his elbow. The other counselor, also a strapping young man, had a wounded leg, and walked with a bad limp. They had both been honorably discharged, even though the war was still on.

That summer, I heard *real* war stories. Our counselors told us how they would have to march and march for days without sleep, and how they might be in the middle of a battle and have to go to the bathroom while they were running. They gave us explicit, uncomfortable details like that, and described how their friends were killed. War suddenly lost its glory and I gained a shocked understanding of the gore and discomfort. I also realized ordinary men were capable of extraordinary acts of valor.

CHAPTER 4

Growing Up in the Bronx

I started public school in the Bronx. By fourth grade I already had political aspirations and ran for the class president. It was a dead-on heat with half the class voting for me and half the class voting for my opponent. I determined that the deciding vote would be cast by Freddy who only sat one seat away from me and was absent that day. I had to figure out how I was going to get to Freddy.

Freddy loved Superman comic books, so the next day I brought in a Superman comic book. Before the class even started or Freddy knew anything about the election, I gave him the comic book. I just said, "Freddy, I'd like you to have this." Nothing else. I figured if Freddy was the kind of guy I thought he was, he would vote for me. He did. I became class president with my first act of political savvy, kind of bribing Freddy.

When I was in the fourth grade I took an IQ test. My parents were informed that my score of 156 was the second highest in the school, and a rapid advance program was recommended. That meant I'd go through the next grades in an accelerated fashion, and continue to accelerate until I was ahead by a whole year. The only trouble was that I was already a good six to eight months young for my class. When I finally skipped ahead an extra year, I was almost a year and a half younger than my classmates. Even a year makes a big difference in a young boy's social development, so basically I was socially retarded compared to the others.

In school, most of my good friends were a year behind me, but I didn't fully realize the difference between me and my older classmates, and the impact that could have until I got to junior high school. I had other problems there, too, which I will explain later. However, my experiences served as a lesson for my own parenting and I wouldn't allow any of my sons to be pushed too far ahead in school; we kept them in their own age groups.

At times, especially when I knew there was a test in school and I wasn't prepared, I got very nervous. When I got nervous, my hands dripped with sweat. This was in the days when we used fountain pens with real ink, so when I wrote the exam, the ink would get all over my hands and run down the paper. Sometimes I faked being sick, so I could escape having to go through this embarrassment.

Concerned about my health, my mother came into my room to take my temperature. My bed in the Bronx was next to a radiator, so I would hold the thermometer against the radiator as soon as she stepped out of the room. That tactic worked three or four times, until one time I overdid it and the thermometer read 106 degrees. My mother went to the store to buy a rectal thermometer, then stayed in the room while she took my temperature. I never got away with that trick again.

When I was sick, my mother brought me treats: jars of melon balls and maraschino cherries, and little cakes with orange frosting from Horn and Hardart. (Horn & Hardart was a big chain cafeteria, popular in the Depression. In those days, 10 cents bought a sandwich or macaroni and cheese there.)

Another treat I loved as a kid was Necco candy (little colored sugar wafers). And my absolute favorite was Charlotte Russe, whipped cream on top of a very thin piece of pound cake in a cup. My mouth waters just thinking about it.

At about ten years old we were curious about cigarettes. My mother smoked cigarettes, and we had seen kids a couple of years older smoking, so Stanley and I decided to try it for ourselves and see what it was like. The mail room in the lobby of the apartment house was kind of hidden, so we chose to smoke our cigarettes there. What we didn't know was the room had an open grill, and all our cigarette smoke was going out the grill. People quickly figured out what was going on, and must have told my mother. When I got upstairs, she didn't holler at me. Instead, she took it surprisingly calmly, saying, "I understand you like smoking cigarettes now."

"Well, I, I . . . ." I stammered, knowing I'd get in more trouble if I lied. I came clean and admitted that I'd tried smoking.

"It's okay for you to smoke," she said, "but smoke in the apartment. Here, let me show how you do it correctly." She lit up a cigarette, then had me take a puff. "Inhale it," she instructed. "Try to blow it out through your mouth or through your nose, but suck it in."

I did that twice. My head started to spin. I ran to the bathroom and threw up. That was the last time a cigarette ever touched my lips for the rest of my life.

My very first hero was a Jewish baseball player and oddly enough for me, he wasn't a Brooklyn Dodger. When I was a child growing up in the Bronx in the late thirties and early forties, the talk and athletic hero of the Jewish community was Hank Greenberg. My father spoke about him all of the time because he too was a Jewish boy from the Bronx, whose parents were from Romania and he had gone to NYU, the same year as my father.

Jewish ballplayers were a rarity in the first place. They never used their real names so people wouldn't know they were Jewish and they wouldn't have to take the abuse that Greenberg did as a ball player. He was the very first to use his real name and to acknowledge that he was a Jew. He was and still is probably the greatest right hand hitter of all times. In 1938, he set a record for the most home runs by a right-handed hitter. When Greenberg hit his 58th home run, with only five games left in the season, he needed three more to break Babe Ruth's record of 60 homers. The pitchers kept walking him or hitting him rather than allow a Jewish ballplayer to break the record held by Ruth.

The Tigers were in the World Series one year with Yom Kippur approaching. To compound matters, the entire Tiger infield hadn't missed a game all year and were about to set a record. When Yom Kippur came, Hank refused to play. The Tigers lost the game, but Hank had won the city's respect. When war was declared against Japan, he was the first ballplayer to enlist. Greenberg became the first Jewish ballplayer to be inducted into the Baseball Hall of Fame, later to be joined by the only other one, Sandy Koufax, an early childhood friend.

Many years later, an interesting thing happened regarding one of the four girls I grew up with in the Bronx apartment. At that age my friends and I wanted nothing to do with girls, so I never talked to one particular girl in the six years we lived in the same apartment house in the Bronx. I only remember Reddie as a tall, skinny kid with glasses in my grade at school.

When my grandson was about five or six, we took him to a computer store specializing in educational toys for children, Noodle Kidoodle. My wife had just taken him into the store while I was out in the limo on a business call. My grandson met a friend inside the store who called out to him, "Ben! Ben Rechler!" His friend was also with his grandmother, and when she heard the name Rechler, she spoke to my wife. Judy brought her out to the car.

Now, here she was, some fifty-plus years later, sticking her head in the car while I was on the phone. "Do you know who I am?" she asked.

For a second, I hadn't a clue who she was, then it suddenly came to me "Reddie?" She was astounded that I remembered her name, and I was too. Judy couldn't get over it either; usually I can't remember what I ate yesterday.

My two best friends in the Bronx, Stanley and Walter, were interesting. Stanley and his family lived directly across the hall. His parents were from Russia, and every Friday night his mother made *verenikis*. It was like a perogy, a dumpling made of soft dough, filled with potatoes, slathered with butter and topped with sour cream. I went there for dinner on Friday nights because I loved those *verenikis*.

Stanley's father worked as a furrier, but didn't have his own shop. No matter what the news of the day was, he would listen to how the Russians were doing and give an oratory on the Labor Party and Socialism. He read the *Daily Worker* (Communist newspaper). I didn't think much of it because it didn't have a sports section. At the dinner table, he discussed unions with Stanley's mother, making it hard for anyone to talk about anything else, but I put up with it because I loved the food so much.

In 1946, Walter's family was the first in the neighborhood to get a television. Up until then, the only TV we ever saw sat in the window of the radio store on our street corner. Sometimes, we'd stand with the crowd outside the store window to watch a big fight on television. After *Wally Gator's* family got their TV I was invited to his apartment every Tuesday night for dinner and to watch Milton Berle. Grownups were invited, too. The Hirsches became very popular people and often had a crowd over. Chairs were lined up in front of the TV like in a theater. My family didn't get a TV until after we'd left the Bronx, so it was a real novelty then.

When we lived in the Bronx I had a few fights in public school. Most of them happened in schoolyards, or when we were going to the movies; sometimes the bigger kids would try to take my money.

I lived in a neighborhood where my side of the block was basically Jewish. Mostly Irish lived in the two new apartment houses across the street. As you'd leave the neighborhood, different blocks attracted different ethnic groups. By the time I was old enough to go to junior high school, I might have to go through two different ethnic neighborhoods, like Irish and Italian, to get to school. Once or twice a week, I had to fight my way through them.

Street fights then weren't like they are today. We never used anything but our fists, and the numbers were fair, so it was honorable that way. If you walked into a neighborhood and some kids picked on you or pushed you, or if you challenged one kid, their leader might fight you. Even if you won, the others wouldn't jump on you. They might break up the fight, but they wouldn't jump on you, so there was a certain unspoken code of honor.

If two kids got into an argument, they might not actually fight each other. One kid would say, "You wanna fight? I'm going bring a friend for you to fight." Then the other kid says, "I'm going to bring a friend." Usually, I was the friend, so I got into a few fights with people when I had no idea what I was fighting about.

I didn't really like fighting and I never started fights. I fought because of injustice; somebody got picked on or someone called me a name, so I stood up for my own honor or someone else's. Sometimes it was a religious slur or just a name like *Fatso* because I had gotten chubby, which was not surprising being fed by all those grandmothers.

One time, when I was twelve, I spotted a crowd in the corner of the schoolyard and went over to investigate. Some rough-looking older kids who didn't go to our school had made Stanley take his pants down and he was in his underwear; they were laughing and taunting him. Furious, I stepped in and demanded, "Let him go."

Now, I should have known better. This wasn't the usual neighborhood group and I got myself into a fight with someone much larger than me. Being worked up about the injustice, I had so much adrenaline pumping that I hit him pretty hard and hurt him.

When I hit him again, he said, "Hold it." Like a moron, I stopped. He walked back to his friends. Suddenly, he came back and hit me full in the mouth. He had put on brass knuckles, and his punch broke my tooth.

My mother was none too thrilled when I came home with a broken tooth. I told her that I ran into a fence because I didn't want to tell her that someone hit me in the mouth with brass knuckles. She'd have locked me in the house for a week for getting into another fight, and learning that someone had used brass knuckles on me would have terrified her.

She took me to a dental specialist on Fordham Road in the Bronx, which wasn't nearby. He had to put a cap on the broken tooth, and told my mother it would last until I was 16 or 18. It was finally changed when I was in my fifties. He must have been a great dentist.

Roger started public school when I got to junior high in the Bronx. His school was about three or four blocks away from mine, and every day that Roger was in public school—even in kindergarten—he got beaten up by some group of kids. It was my job to run all the way from the junior high school to the public school and pull the kids off Roger.

I had to teach Roger how to fight, but he was afraid and lacked confidence. Telling him I had a secret formula, just like the comic book heroes used, I mixed up some purple goop with my chemistry set. "Rub this on your arm," I told him, "and all you have to do is say 'purple paint' and you'll be able to beat anyone, even if they're bigger than you."

That made Roger almost fearless. He started to fight better because he wasn't as frightened and he didn't get into as many fights to begin with. However, one time he went to camp and encountered a bully twice his size. All the counselors hated this tough kid that was in Roger's group. One afternoon, the camp held a boxing match and poor Roger went into the ring with the bully. "Purple paint, purple paint," Roger repeated under his breath. Needless to say, he got the living daylights beat out of him. He's been mad at me ever since.

When we weren't at camp, Dad took us on vacations in the summers. He loved to let loose and have a good time on vacations, as he did at parties. On vacations, he always brought a movie camera along and loved to do funny things he'd seen in the movies, like act like Charlie Chaplin and make us laugh. Inevitably, he'd come home from vacations with a limp or a cut on his forehead from clowning around on the diving board or while playing with us.

One time, he bought my brother a ping-pong gun (a handgun that shot a ping-pong ball), and had me take movies of the two of them. He put an apple on Roger's head, wanting to shoot it off like Errol Flynn. Instead he shot my brother between the eyes. Roger screamed hysterically (he was seven and the ping-pong ball scared him). In order to pacify Roger, my father put the apple on my head and let Roger shoot at it while my Dad took the movie. Naturally, Roger missed the apple and got me. That made Roger feel better, and it was my father's way of rectifying the situation.

One time, he took us to Jamaica, and I recall two memorable things about that trip. First, the day we got there, my father bought a Cuban cigar, proud that he'd only paid seventy cents for it. In the fifties, seventy cents was like three or four dollars today, but Cuban cigars cost much more money. When we were two blocks away from the cigar seller, he realized it hadn't cost seventy cents, but seven dollars. He'd made a mistake with the exchange. I've never seen him run so fast as he did those two blocks to get his seven dollars back. He got it back.

The next incident involved a very colorful wagon pulled by a couple of donkeys and filled with watermelons. My father went out in the middle of the street and started to take a movie of the old man with the big straw hat driving the colorful wagon. The man waved his hand as if to say "no."   Not knowing if the man was waving at him or not, my father kept on taking the picture. The old man jumped off the wagon with a machete, and chased my father down the street. Later, we found out that in some cultures, they believe that every time their picture is taken, it takes days off their life. Older people, in particular, still believe in those things.

When we weren't vacationing, my father always sent us somewhere in the summers to escape the heat in the city. At first, we went to the Rockaways, then we went to the country where we stayed at the Hotel Plaza.

During the war years, my father rented a house right on the ocean in Atlantic Beach. He commuted from the Bronx to join us on weekends, but often didn't arrive until Saturday evening. For two or three summers, we had this house in Atlantic Beach. It was very rural, but there was a candy store within walking distance where I could get my newspapers every morning to check the baseball statistics. Roger and I didn't have any children our age to play with there, but we played stoopball on the big stoop, just the two of us. We could still listen to the ball games together on the radio, and there was an empty lot next door where we could play.

The candy store also sold sexy French postcards, big cards with old fashioned black and white photographs. I started to collect those too, but I didn't bring them into the house. I kept them hidden in a box buried in the overgrown lot next door. I put a rock on top to mark the spot, so I could dig up my treasure when we returned the following summer. But when we returned, I discovered a house under construction next door and my buried treasure was gone. I realized then that you can't count on anything remaining the same.

I never had a birthday party. Today kids have birthday parties all the time, but in those days, we didn't have them. You would take a few friends to a movie or ballgame and an ice cream soda. I got a lot of toys, but these, especially during the Depression and war years, were mostly handmade. Uncles or friends of my parents would give me different articles, such as a real old fashioned fireman's hat, or a gas mask from World War I. My grandmother made me a soldier suit. My father made me a game to roll marbles in that was made out of a cheese box. Because that was all we had to play with during the Depression and World War II, we used our imagination more than kids today.

My father used his ingenuity and creativity to solve his business problems, and I leaned in that direction also. The first example of this happened during the linoleum gun wars we used to have with the kids across the street. You would cut the corner of an orange crate, and stretch a rubber band across it to make a combination gun barrel and sling shot. Then you'd put a one inch square of linoleum, which was hard like asphalt tile, under the rubber band. When you released the rubber band, you'd send the cut of linoleum across the street as a missile and hit somebody. No one ever got really hurt, but they might have—we were standing on opposite sides of the street shooting at each other. In the wintertime we had the same street wars but with snow and ice balls instead of pieces of linoleum.

One time in the midst of one of our wars, they had more kids on their side of the block than we did, so they had more firing power than we did. I figured out a way we could equal them and gain an advantage. I created a three-barrel linoleum gun, so we could get off three shots to every one of theirs. I was only eight or nine when I did this, and looking back now, I realize it was probably the first time I used this kind of creative ingenuity to solve a problem. Creative problem solving was a talent I continued developing for the rest of my life.

Not long before we left the Bronx, I was thirteen and about to be Bar Mitzvahed. For my Bar Mitzvah present my father gave me four box seats at a New York Yankee/Brooklyn Dodger football game. (They had football teams at that time: a Brooklyn team and a New York Yankee team played in one league—American Football Conference, AFC. The Giants played in the other league—NFC.) I took my three friends: Wally, Stanley, and Teddy. This was the first game we attended totally alone. No one put us on the subway or brought us to the stadium.

We saw the game, and it was fabulous! In those days, the fans walked onto the field with the players after the game, so we did, thinking we were going to get autographs. When we walked up to these players we didn't have the heart to ask them for autographs because we could see how beat up they were. I realized then what a tough game professional football really was by how beat up and bloody these guys looked.

When we left the stadium, somehow I got separated from my three friends. I finally heard them hollering at me from across the street. Relieved, I started running across to meet them. Just then, a car rounded the corner. The woman driving wasn't going fast, but she hit me. I must have been thrown up in the air. I blacked out and don't remember anything. I woke up to see a crowd standing over me. I could hardly move. There was a policeman peering at me, so I must have been out for a few minutes. Bleary-eyed, I watched the activity swirling around me. The woman screamed at me: "How dare you run in front of my car like that!" The policeman told me he was calling an ambulance, to stay calm.

"I don't need an ambulance," I heard myself say.

"You sure?" he asked doubtfully.

Somehow I managed to scramble to my feet, insisting I was fine, and he let me go.

I wasn't fine. My whole body hurt and I had a huge lump on my head. When I got home and undressed, I knew I couldn't *ever* let my parents see me like this. I didn't know how I was going to get around the apartment without them seeing my bruises, but I knew I had to. I was black and blue from my feet all the way up to my head. I knew I couldn't tell my mother or I would never have been allowed to go to another game in my life.

I had a slight limp for a while. In about two days my right eye got very small. I didn't really understand what happened, but I must have had a concussion. From that point on, I began to have trouble with schoolwork, and started to behave irrationally. Because I had skipped a year, I was already in the eighth grade.

My mother started taking me to specialists to see why my eye was small, first eye specialists, then neurologists; I must have seen five or six doctors. I never told her or the doctors about the accident. My schoolwork got pretty bad, but I made it through junior high, probably because I was already so far ahead.

Tremont Temple on the Grand Concourse was one of the most Reformed temples in the country at the time. Its Rabbi, Rabbi Rosenberg, was one of the leaders of the Reform Movement. Almost banned for being so vocal, Rabbi Rosenberg spoke out against Israel as an independent state, and even rode in a parade down the Concourse with Franklin Roosevelt. In 1947, when I had my Bar Mitzvah, the war was over and the topic of Israel was being hotly debated in the United Nations. Rabbi Rosenberg felt that after the Holocaust, if the Jewish people had their own country, other countries would demand they leave. He felt strongly that Judaism was a religion, not a country. He thought Israel would become a target for the whole world if the Jews had an independent country, and as history would reveal, he wasn't too far wrong.

I studied with a cantor for about a year to prepare for my Bar Mitzvah. I didn't do the whole services like they do now. I was to have a Bar Mitzvah party at our apartment, but I was only allowed to invite two friends, besides my cousins. It was the grownups' world. My father bought me my first suit at Barney's (Husky Department).

My parents were reformed and not very religious. Their former synagogue, Tremont Temple, had provided very little hebrew. We didn't hold many special ceremonies in our home. Grandpa Rechler ran the services when we went over to my grandparents' house for Passover or such, and he was a classic comic. He ran through the service as fast as possible so we could get to the food. I learned my services from my grandfather, he only did the abridged version, which is what I do today in our home.

CHAPTER 5

High School Years
in the Country

My parents used my graduation from junior high as an opportunity to move. Later that year, we moved to Beechhurst, Queens, in an area called Robin's Wood, which was near Whitestone, Long Island. It was the country. There were woods across the street from our house.

My father had bought a house with a lot next to it, so we had real space, backyard space like we had never known before. My parents loved living in Beechhurst. As I explained, they were both arty people, and by this time my father had done so well in business that they had enough money to own their own home and do their own thing with it. It was a wonderful house, just two blocks from Long Island Sound. My brother and I shared a very long room.

Not only were we the only Jewish family in a one mile radius, but there were few other children, only one boy my age. Having grown close to my friends in the Bronx, I missed having the big group to play with. I now made friends with Dick Godfrey, the one boy who lived nearby, and another boy who lived several blocks away, Roger Willen. In apartment housing everybody lives close, and there are a lot of kids on one block, but here there weren't enough kids to play ball. And at Robin's Wood there was no candy store.

Bayside High School was about five miles away, and there were no school buses in those days, and no public transportation in Beechhurst. As a boy of thirteen or fourteen, the only way to get around was by bike. I rode to school every day with Dick Godfrey and Roger Willen. Along a narrow, dirt path, we peddled across field after field as quail and partridge fluttered up out of our way. It was pretty, and I liked living in the country, but it was *very* different from Brooklyn and the Bronx. In winter, we walked or rode our bikes to school when there was no snow on the ground. As we got older, we hitchhiked.

I still missed my Bronx friends. I'd made such a fuss when we were moving that my parents promised to allow me to go back and see my old friends in the Bronx, but I had to take two buses, a subway and a trolley car to get there. I did it a few times and played ball with them again, but after a year or two, it just seemed too far away and eventually I lost touch with them.

I was a joker when I was a child and enjoyed making people laugh. As I got into my early teens and had more money, I loved to go to specialty stores to buy the latest tricks. My mother was my favorite victim. When we moved to the house in Beechhurst, we had a staircase for the first time. I did a pratfall (faked it) down the stairs. I fell with a thud, and my mother, who was in the kitchen close to the bottom of the stairs, couldn't help hearing it.

As she ran out of the kitchen, I held my stomach and cried, "My kitneys, my kitneys." As she came over to see what was wrong with me, a little kitten stuck its head out from under my shirt. These are the kind of jokes I enjoyed playing. I can see why she didn't know whether to laugh or cry.

My friend, Dick Godfrey, was very much a loner. He didn't play basketball or baseball, but he had a sailboat. By the second year we lived in Beechhurst, my father bought me a sailboat like Dick's, so I'd have something fun to do in the summer. I learned to sail, and even raced some. My sailboat was a 16-foot Comet. (Comets are extinct today, but there are still some Lightnings around).

I'd spend all spring and fall, whenever I could be outdoors, working on the boat, figuring out how to make it lighter so it would go faster. I'd buy the finest fixtures in the hardware store and the marine store. I'd put in a huge amount of time sanding the boat and the mast. My boat was called *The Wild Goose* after a song by Vaughn Monroe: "My Heart Goes Where The Wild Goose Goes." I guess I liked having my own sailboat . . . . I really liked trying to win a race, but I don't know if I liked doing all the work.

The Long Island Sound and the mooring where I kept my boat was just two blocks away. Roger used to help me work on it, and we'd use a dinghy to get from the dock out to the boat. One day, I gave Roger some green paint and he painted the bottom of the boat because he was little and could get underneath. When he came out from under the boat, he looked like a Martian—green from top to bottom, including his hair! My mother didn't appreciate it too much, and as she scrubbed Roger's skin with a brush and turpentine, he wasn't too happy either.

My father was looking for a synagogue to become involved with because now it was my brother's turn to be Bar Mitzvahed. There were no synagogues in the area, so with a group of families in the Whitestone area—with whom my parents eventually became lifelong friends—they gathered enough people together to form a congregation and created the Whitestone Synagogue. At first, they rented an old house and then eventually built a synagogue about fifteen blocks away from where we lived. Roger was one of the first to be Bar Mitzvahed there.

My mother started a Cub Scout group because there weren't a lot of kids Roger's age in the area. An unusual Den Mother, Gloria Rechler came up with ideas that parents probably had difficulty believing when their kids came home to tell them about their projects. Roger had a pet cemetery in the woods. He was the undertaker of the neighborhood. Whether it was a cat that got run over, a squirrel that fell out of a tree, or a bird that flew into a window, Roger picked it up, and, with a group of friends, buried it in his pet cemetery.

In my first year of high school, I continued to behave irrationally and still exhibited that small eye that manifested itself after the accident. If a teacher sent me to the blackboard to do a math problem, I might just as soon draw a picture. With my paperwork I would do the same thing, or I'd hand it in blank. I failed two courses that year, and my parents were beside themselves trying to understand why I was doing so badly. I'd had decent grades before my concussion, which they still didn't know anything about. Although I'd been called an underachiever, schoolwork had always come easily to me. Not anymore.

Life can be a struggle for any teenage boy, but at this same time, my life had completely changed. I'd had to change schools *and* friends. Living in the country was very different from living in the city. People were different. There were no more fights, no groups of friends. I struggled in many ways.

After I failed French and geometry, my parents hired tutors. I gave up French because it was just too difficult for me, and transferred to Spanish instead. I studied Spanish all that summer with a tutor before I took it in school. I also went to summer school at Jamaica High to raise my "D" in geometry, and got 85% on the final exam.

My summer from hell continued into the fall. Regents Tests were administered in New York schools now, and if you didn't pass the Regents you couldn't get into college. When I returned to school for my sophomore year I was put in a general education class, a non-Regents class. Even if you graduated from this class, you wouldn't be eligible to go to college. I was afraid to tell my mother about it at first, and at the same time I didn't care much about the thought of pursuing college.

From the first day of school I realized I didn't belong in that class. It was like "Blackboard Jungle." The kids were wild and unruly, and really, really dumb. An ex-Army captain was the proctor, and I use the word proctor rather than teacher because he taught very little. He just tried to keep order, poor man. Every time he turned his back, missiles went flying all over the room. It was almost like a TV sitcom. I knew I didn't belong there, and had to get out.

After about three days, the proctor said to me, "What are you doing here? You don't belong here." I finally told my mother, and she came to the school. It turned out that they had never given me credits for the courses I'd taken in summer school. They re-tested me in geometry, and I scored 92% this time. I did just as well on the Regents when I was allowed to take it. From then on, I had tutors all the time. I had trouble with English, particularly spelling, and may have been somewhat dyslexic. In math, the answers came naturally.

There was a big improvement in my junior year. I understood that I had to work hard to go to college because good grades wouldn't just materialize. In earlier grades, whenever I took a test my hands sweat profusely, making the ink run on the paper. Now I realized if I prepared and knew my work, I didn't have to worry. All I could do was my best, and if that wasn't good enough, so be it. From the moment I came to that realization, my hands stopped sweating. Throughout my life I have always tried to be prepared, albeit sometimes I prepare at the last minute, but I am usually prepared. It was a valuable lesson.

Later, as a young father drawing on my own school experiences, I realized that education is not the means to an end, it's a tool to help you get there.

The summer I was about to go into my junior year, we finally put a large enough group together to play baseball. We used to play right on the Cross Island Parkway in the triangle between the exit and the parkway on an uphill slope until the park police would chase us off. Every once in a while traffic backed up or we would hit a car, or the ball would go inside a car. I remember once climbing into a convertible to get my ball, while a woman screamed at me.

When I was 15 and had filled out physically I also started to play golf. Sometimes we would get up and play nine holes before school, then play another nine holes at night. I played a lot of golf after graduation, the year I worked at Rexart before college. I would play at Clearview Golf Course before work and again after I got off.

Along the way I'd played softball in an industrial league with a team from my father's factory, but when I was 15, I finally found a real baseball game. Sylvania, the light bulb company, had a regulation baseball field behind their headquarters building on Cross Island near Clearview Golf Course. No one else from my group went. I was the only one good enough to play baseball with the employees there. The Sylvania people would say that I was the son of one of their employees so I'd be eligible to play, but I was really a ringer.

In 1950, at the age of 15, I pitched for Sylvania. I would pitch almost three-quarter sidearm, almost underhand, and I threw the ball very hard. I emulated one of the best pitchers of the times, Ewell Blackwell. I played with them for two years in the spring and summers when I wasn't working. I had a very live arm, even though I was wild and most of them never got a hit off me. I walked a lot of players who were afraid to stand up because the ball came so fast and so wild. Eventually, my arm went because I'd thrown so hard at such a young age. The only one that enjoyed it more than me was my black cocker spaniel, Major. He would park himself under the beer keg and catch the drippings.

My junior year was good, and I breezed through my senior year. In my last two years, I had worked after school at my father's new business, Rexart. The plant was a ten-minute bike ride from my house. After graduation I worked there almost full time. My parents held me back from college for about a year because they felt I was too young at 16 to be away from home. The next semester I took two physics courses and pre-college math because I had decided to go to engineering school. My goal was to become an engineer, then return to

Rexart and make an important contribution to my father's business.

My mother felt I had no capacity for business and would not make a good businessman. She suggested that a college professor would be a good career fit for me; it would give me a life that I could handle because professors didn't work too hard. It didn't seem to matter to her what I was actually *interested* in doing.

One day, when I was riding my bike to work at Rexart after school, I got pulled off the road by some members of a heavy union that was trying to unionize Rexart. Uncle Mort had hired all veterans, tough in their own right and disinterested in unions. That day, the union brought a busload of hoods who looked as if they'd just stepped out of a gangster movie. They wore suits with fedoras and carried baseball bats and steel knuckles. If anyone tried to drive in to work, they would smash their windshield. My father and uncle saw what was happening and called the police, but not before the hoods grabbed me off my bike and threw me against a fence. They must have thought I was a worker. Well, I was, but I was also a kid on a bicycle. A few of the employees got close to me and I could tell they were trying to protect me, but no one knew that I was *really* one of the bosses' sons. All of a sudden, I heard police sirens. The gangster-types hustled back to the bus and took off fast.

Rexart ultimately was to be unionized, and that was done by Harry Davidoff, a famous Murder Incorporated suspect the authorities could never convict. In later years, he ended up running the JFK Airport's union and was allegedly involved in many drug and theft scandals at the airport.

During the summer of my junior year as well as several months before college, I worked at Rexart full-time as a clerk in the shipping department. Once, when I was loading boxes of chairs onto a truck, I got my arm caught in a moving conveyor belt. The conveyor belt had two steel rollers and each one had a belt wrapped around it. I loaded a box onto it with my hand under the box, and my hand got caught in between the two rollers. Unable to reach the controls in time to stop it, my arm was pulled all the way up . . . and kept going in. I struggled, but couldn't pull my arm out. They had to run the conveyor belt in reverse to back my arm out. I have strong cartilage and bones (like fish bones). I never once broke a bone. When I finally pulled it out, my hand and half my arm were compressed and seemed about a quarter inch thick.

Someone ran to get my father and he took me to a nearby family doctor, a European gentleman. The doctor looked at my arm, then x-rayed it. "I don't know how you did it," he said, examining the lighted x-ray picture, "but you're going to be okay. You didn't break anything." My arm returned to normalcy; it is still good today and I've never had a problem with it.

By the time Dad and I drove back to the factory that day, my hand had swollen up to the size of a balloon. He was petrified to take me home to my mother, so he took me back to the factory to call her and prepare her first. When we got back to the shipping department, there was blood all over the place. "Did you bleed?" my father asked, alarm showing on his face.

"No, I didn't bleed," I assured him.

"Then, what the . . . !"

All of sudden, the shipping clerk ran up to us. It turned out the truck driver had tried to show him what had happened to me and had caught his arm in the same conveyor belt. He crushed every bone, and finally lost his whole arm in the hospital. I was very lucky.

Growing up, I saw very little of my father because he worked six days a week until 10:00 PM. He would come into my room before he went to work and wake me to say goodbye. He'd read me stories from a history book while I was in bed.

On Sunday mornings, I often lay on the floor quietly in front of the radio and listened to Mayor LaGuardia read the Sunday comic strips to children. When my father got up, he'd make breakfast for me (and Roger after he was born), then bring my mother her breakfast in bed. It was always some concoction, some invention of his. He'd get creative, and more times than not, it missed—like the time he made scrambled eggs with garlic cheese.

My mother was very much ahead of her time. She read all the latest articles and kept up with the newest trends, and every night, she listened to the first talk show on the radio—Barry Gray. She also listened to Carlton Fredricks, who was the father of natural medicine. If there was anything the slightest bit wrong with me she took me to the fanciest doctors. In retrospect, I realized she probably did that because her mother had taken her to a Park Avenue doctor when she had polio and she had survived it. She really investigated the newest trends in medicine and she'd drag me all over the city to see specialists long before it was fashionable. Some of the newer remedies weren't always good, but she was the first one to try them.

Beginning my senior year, I had a bad case of acne. My mother heard of a doctor who gave x-rays–radiation–on the face as an apparently successful treatment. Willing to try the latest in medicine, my mother decided I should take these treatments. As we later found out, radiation treatments were not without peril. For the most part, they worked on the acne, but that doctor's fingers turned black and he eventually lost them and then died.

This kind of treatment was too far ahead of its time, and they just did it without knowing the end results. Years later, I had a small cancer removed from my face, and it occurred to me that the cause may have stemmed from those early radiation treatments.

In 1951, while I was in high school, the Brooklyn Dodgers and the New York Giants were tied in the World Series playoffs. After class, I ran all the way home to see the last two or three innings of the deciding game. People rated that game as the number one baseball game of all time. It's called *"the shot that was heard around the world."*

In the bottom of the ninth, there were two men on, two outs. The bases were loaded. Bobby Thompson got up. To my shock and dismay, the Giants managed to snatch victory out of the mouth of defeat! Thompson hit a home run right down the line, a pop fly, 290 feet into the stands. The Dodgers, the superior team, lost the pennant in 1951. All that was left was Brooklyn's familiar refrain: *"Wait till next year!"* That game was the biggest heartbreak in my lifelong love affair with sports.

There is nothing like watching the really great teams. The Dodgers of the early 1940s continued to be great after the war when many of their original players came back. Through the late 1940s to 1956 when they broke my heart and left Brooklyn, they fielded sensational teams, even though the damn Yankees beat them most of the time, until the very last year before they left when the Dodgers won the World Series in

1957. After that, they moved to Los Angeles and I felt as betrayed as the rest of the Dodger fans—like we'd just gotten a Dear John letter.

In 1947, my favorite ball player was Jackie Robinson, the only black player in organized baseball then. He was a great player, but not the best player of his times, probably not even the best on that great Dodger team, but there was something special about him that even a twelve-year-old could sense. He was heroic and one of the most exciting players to watch, especially the way he would steal home. He wasn't the best hitter or fielder, but he just had a knack for making the most important play at the most important time. He was a winner. When the Dodgers finally moved out, they tried to trade Robinson to the Giants and, typical of him, he refused to go to the hated enemy, and left baseball.

Signed by the Brooklyn Dodgers, Branch Rickey, the owner, after careful consideration, picked Robinson to break baseball's segregation. It was done amongst much protest. Up to that time, black players were only allowed to play in the Negro League. Robinson had graduated UCLA and was an All-American in four sports: football, baseball, track and basketball. He had been an officer in the Army and was exceptionally bright. Knowing that Robinson would be the target of others' ignorance, Branch Rickey had Robinson promise him that no matter how much abuse he would receive, including threats on his life, he would not retaliate. Robinson was, by nature, one of the most fiery, competitive athletes ever to play any sport. In his first year, 1947, when I was only twelve, it was apparent to me that he was not only a great ball player but his heroism transcended the baseline. There was something special about him, even heroic in how he endured the abuse and even deflected it by competing harder. He was by far the most exciting ball player that I have ever seen and he was a precursor of the Civil Rights Movement. He was inducted into the Baseball Hall of Fame and recently his Number 42 was retired not only from the Dodgers but from all of baseball and is hung on the outfield walls in his honor. He has remained the symbol for integration not only for baseball, but in all walks of life. Roger Kahn, a famous sports journalist wrote, "Robinson had intimidating skills and burned a dark fire. He wanted passionately to win. He bore the burden of a pioneer and the weight made him more strong. If one can be certain of anything in baseball, it is that we shall not look upon his like again."

The Giants had a great football team in the 1950s and into the early 60s when they had such players as Frank Gifford, Alex Webster, Y.A. Tittle, Del Shofner and such linemen as Roosevelt Brown and Robestilli. They won a couple of championships, but then they went through a terrible stretch when they were one of the worst teams in football, coming back finally to win the Super Bowl in 1987 and again in 1994. I bought season tickets for the Giants games when I got out of college in 1956 and I still have them, almost 50 years later.

The summer before I went away to college, my parents invited me along on their vacation to Taxco, Mexico for my graduation present while Roger was in camp. Taxco was the sterling capital of Mexico where local silversmiths crafted lovely, modern and artistic pieces out of silver. It was a mountainous area, and we stayed in a long, single-story building. My parents occupied a room on one end and I was at the other.

On our second day, I saw a very pretty Mexican girl at the newsstand, the niece of the owner. Being incredibly shy at the time, I just hovered around the newsstand and stared at her. Finally, she noticed me and started to laugh, then spoke to me in English. I went into that newsstand four times in the next two days to talk to her.

One night, my parents went to dinner, and I chose not to go with them. I ordered in room service, and that same girl brought the food to my room. Well, that was my first real exploration into the other sex. I was sixteen and about to go to college, and I would guess she was fifteen or sixteen. I remember her clearly: she was petite, dark-haired, dark-eyed and very cute. This was my first heavy necking experience, and she was the one who initiated it. I was so shy with girls that I didn't even know how to talk to them, let alone initiate anything else.

I thought I was in love. We had to leave the next day, and I was heartbroken. However, the experience bolstered my confidence.

"Even a blind squirrel finds an acorn once in a while." — Donaldism

CHAPTER 6

College, Fraternity

and Intramurals

I applied to three good engineering schools: Clarkson, Lehigh, and the University of Miami. I got accepted by Clarkson and Miami. Clarkson was right on the Canadian border, so I chose to go south to Florida because of the weather.

The University of Miami accepted a freshman class of 2600 that year, but approximately only 200 of my original class graduated. (There were 362 graduates in my class, but only 200 had started as freshmen and the rest had transferred in.) Like any serious educational pursuit, whatever you put into it you'll get out of it.

The truth was that you could get an excellent education at the University of Miami. Many highly respected professors who were retired or near retirement took tenure in Florida and just taught a few courses at U of M. I had some first-rate professors in my years there. But Miami students had a reputation. People would say: "What are you majoring in? Underwater basket weaving?"

In my freshman year, I attended a packed psychology lecture in a hall with about 300 students. Professor Fischer kept the session open to questions, but he was such an unbelievable lecturer and covered the material so well that he got very few questions. Fischer was famous because he'd solved a major serial murder mystery in Chicago at that time. This was the caliber of professors we had. Later on, when I changed my major to industrial management and time and motion study, two of the people who wrote the books that set the trade standards were my professors at Miami. One was Dr. Lesperance and the other was Gilbreth, about whom a book has been written—*Cheaper By The Dozen.*

When I arrived at college, I was assigned to a dorm. More like an apartment, the dorm had two bedrooms, a living room and a kitchen, and I had three roommates. One of them, a very large guy, was simply nuts. Every day he would switch beds; he'd take my bed one day, and then the next day he'd want the other one. Finally, I told him I refused to move anymore, and he started to choke me. It took the other two roommates to pull him off. Thankfully, the nutcase only lasted for three weeks before he dropped out of college. My other two roommates were very nice.

I decided to rush a fraternity; my father had encouraged me because of his own good fraternity experience. Dad had been in a fraternity in high school and his closest friends were still his fraternity brothers. I rushed Phi Sigma Delta, the smallest fraternity, and there I met Bruce Tucker. He was also considering Pi Lam, a larger fraternity. Pi Lam was all about impressing girls and driving red convertibles.

Bruce and I hit it off immediately. We went to all the pledge parties together and I spent a whole night encouraging him to stay with Phi Sig. With only 36-38 members, Phi Sig was a jock fraternity that enjoyed playing a lot of ball, even though most of the fraternity was shorter than my majestic five-feet-eight. Bruce had been a high school football player, an all-around athlete who had received a bonus offer to play baseball from the St. Louis Browns. He'd chosen to go to college in Miami.

Although he was five-feet-six and over 190 pounds at that time, Bruce had a big-boned, athletic body. He had large muscular legs like the running back he was, and no idea of his own strength. It was obvious from the very beginning that Bruce wasn't used to losing. I later learned, that on the rare occasions when he did lose, whether on the ball field or the card table, his eyes would swell and turn red and his body would tighten up. The ultimate competitor, he hated to lose, but even so, he always remained the model of sportsmanship.

Bruce and I pledged Phi Sigma Delta together, and that began our lifelong friendship. He had a good sense of humor and was extremely popular in school, but more importantly to me, Bruce Tucker was a friend who could be counted on.

The times were very different in the mid-fifties. In my freshman year, Bruce and I and three of our rush brothers were having dinner one night when they decided to join a "panty raid" at the girls' dorms. There was a ten-foot fence with a locked gate and a guard outside the girls' dorms. The boys would clamor on the other side of the gate, yelling and waving, and the girls would open up their windows and throw down their bras and undergarments—this was a panty raid.

Not thinking much of the shenanigans one way or another, Bruce and I went to play cards instead. That panty raid turned out to be a major to-do. Three of our pledge brothers were suspended from school because they were picked out from photographs. It seems like such a harmless, ridiculous prank now, but in the 1950s, it was scandalous.

I threw myself wholeheartedly into college life. I had pulled my grades up and had done well in my last year in high school, but I'd never enjoyed being at school the way I did now. I loved the fraternity and the friendships I was making. Unfortunately, it was evident to my father that I was enjoying myself too much when he saw my mid-term grade in accounting. He told me that if I got anything less than a "C" in any of my courses, he would take me out of school.

In that first semester I had trouble with only one course—accounting. I was getting "B's" in all my other courses, but I was seriously struggling to get up to a "C" in accounting. By the finals I still wasn't sure I'd make it, so I was very worried that my father would pull me out of school. This led me to do something stupid.

The finals were held in a very large lecture hall, taken by the entire class at one time. My roommate had a similar problem to mine, but in history. We decided that I would take his history test and he would take my accounting test. It was a big chance to take, and this might have been a very different story if we'd gotten caught. We didn't. We got away with it. Some of my fraternity brothers knew, but no one else ever found out. My roommate and I both felt relieved that we had passed, but doubly relieved that we didn't get caught.

Friday night was the big night for us. The University of Miami always had great football teams, and their home games were on Friday nights. Win or lose, the fraternity had a party after the game. They served this nauseating drink—called Purple Passion—made with grape juice and cheap champagne, with sherbet floating in it. If you drank too much, you got some hell of a headache. I wasn't much of a drinker, but I learned fast not to drink much of this.

One Friday night, I took a date, Myrna Pine, to the big game against Notre Dame. (Her father was the owner of

Hebrew National, the major delicatessen supplier.) To get into the game all you had to do was show your student card; you didn't have to pay any money to get a ticket. Each fraternity had their own section, so there were good seats awaiting us.

All excited about the big game, I arrived at the ticket window with Myrna. She didn't have her student card. Well, I wasn't going to miss this game, and besides, there was a fraternity party afterwards. Myrna didn't have a penny, and I had about 50 cents in my pocket. All that bought was a bleacher ticket. This was Miami in 1952—the Deep South—and the only people sitting in the bleachers had black faces. I didn't want to miss the start and it was too late to go back and get it. I bought the bleacher ticket for her. I sat in the Phi Sig section, and looked over to the bleachers where I saw this one little white face. Needless to say, Myrna and I never dated again.

In November of our freshman year, Bruce and I decided we weren't going home for Thanksgiving because it was only a four-day weekend. At the last minute we changed our minds and decided to go home and surprise our parents. Bruce had grown up in Merrick, on Long Island. We managed to get cheap flights to New York, and boarded the plane just before midnight, thinking what a great surprise it would be for our parents.

I arrived home about 2:30 A.M., and attempted to slip in the back door very quietly. I hadn't even fully opened the door when I heard my mother call down to me, "Donny, take out the garbage," as if I'd never left. Was her intuition at work again?

The second semester of my freshman year, Bruce and I moved off-campus to a frat house. We shared a room with another fraternity brother, Howard Strauss, who had just pledged with us. Rooms were chosen by seniority, so we took what was left at that point. Across the street from a river in Miami, the frat house was a good fifteen to twenty minutes from the campus in Coral Gables.

One night, I was going on a double date with Bruce to the football game. He actually had a red convertible by then—a new Chevy that his father had given him in the second half of his freshman year. Just before the game, his date canceled, but now I had no way of getting to the game. Being the good friend he was, Bruce lent me his car. I wasn't anxious to borrow the new convertible, but I did.

I didn't know how to get to the Orange Bowl because I'd never driven there before. I picked up my date and followed one of my fraternity brothers—I think it was Danny Gordon—to the Orange Bowl. Halfway through Coral Gables, Danny went through a yellow light. Afraid I was going to lose him, I followed him through the light and got broadsided. Nobody was injured, but needless to say I felt awful about the car.

Arriving at the game a half hour late, I had to face my best friend and tell him his red convertible, the gift from his father, was demolished.

Bruce just looked at me with concern. "That's all right. Are you okay?"

We were two eighteen-year-old boys who, even then, knew we were going to be lifelong friends.

The fraternity house was pretty wild and woolly. The two-story house had been a boarding house before we took it over. Unlike college dorms, where someone made you clean up, the frat house was usually a mess. Thankfully, we had cleaning help once a week, so it wasn't a total disaster.

Every day at about three o'clock a tour boat would pass by on the river. One of the brothers with wild-looking red hair that stood up all over his head, would run out on our balcony in his undershorts, yodeling like Tarzan and pounding his chest. We heard the tour guide on the boat's loudspeaker pointing him out as a wild man who lived in the house. My red-haired frat brother became one of the local attractions.

Another funny story involved another brother, Larry Ornstein. He convinced this girl to stay overnight with him at the fraternity house. That didn't happen as often as you might think because this was the 1950s.

The next morning, Bruce and I got up early, before the rest of the house because we were going out for breakfast, and then out to play ball. On a huge mirror in the living room, we saw a message written in lipstick: "Larry, I'm going to kill myself. I'm going to jump in the river and end my life." She felt guilty that she'd stayed overnight.

We ran upstairs so we could see the whole riverfront better. Looking out the window, we spotted the girl standing on the bank of the river like she was going to jump in. Being business students and having just finished Business Law 201, we realized that if we touched her and she drowned, we could be held responsible. We decided against going out to grab her, and ran into Larry's room to wake him up.

Larry jumped into action. He pulled on his pants and ran downstairs in bare feet, carrying a towel. When we got to the living room, he was standing on the couch wiping his name off the message. Finally, a couple of guys went out and talked her into coming back from the river. But the funny thing was that instead of going out to rescue her, Larry had wiped his name off her message.

We had some crazy parties and all night card games. In fact, the card games weren't just one night; they'd start at the beginning of the semester and there would be enough interchanging of people that the game never ended. There was one continuous poker game in our dining room for an entire semester.

Despite the parties, we played a lot of ball. The school had a very large intramural program that included every sport: baseball, basketball, football, boxing, wrestling, tennis, ping-pong—anything you could think of. The teams were made up of different fraternities and clubs and groups, even one or two dormitory teams. Some of these teams were made up of Korean War vets. With the war recently over, a lot of these vets had come to Florida on the G.I. Bill. They were men, and we were boys playing against men.

After my freshman year, I spent the summer back home working at Rexart. My father surprised me by giving me his old Buick. I continued my x-ray treatments for acne. The big event of that summer happened when I took my cousin, Ann, for her interview at Sarah Lawrence.

For some reason, my aunt wanted me to drive her. Growing up, Ann had been treated very much like a débutante by my aunt, and was groomed for Sarah Lawrence. I drove the Buick to New Rochelle where I picked up Ann and drove her to Sarah Lawrence. Parked outside the fence while she went in, I could see into the school window and saw these two little old ladies interviewing Ann.

I turned the motor on to switch on the car radio and listen to the ballgame. It seemed like I sat there for a long time—about 45 minutes. An elderly man watered the flowers along the edge of the fence, taking his time.

All of a sudden, there was an explosion and fire streamed from the engine of my car. I popped open the hood and tried to fan the fire out with my shirt. That didn't work, so I quickly asked the aged gardener for his hose. The water barely dribbled out. "Turn it on!" I yelled.

Instead, the gardener pulled the fire alarm. The little old ladies came running out of the interview with my cousin. The fire department arrived, sirens wailing.

I was accident-prone in those days, some accidents due to myself and some were inadvertent. My aunt had warned my mother, "Tell him to make sure he doesn't do anything. Just sit in the car. Don't get into trouble." Of course, I created a stir. Thank goodness, Ann got into Sarah Lawrence anyway.

That summer, I saw a lot of Bruce and another fraternity brother, Stanley Rabinowitz, who would also become a lifelong friend. Stanley lived in Flushing, one bus ride away from me in Beechhurst. We had to drive further to Merrick to visit Bruce.

A year ahead of us at college, Stanley didn't partake in many of the daytime fraternity activities or sports. During the school semesters, he worked as a salesman in the daytime in the Miami area. His father had a large electric supply business based in New York City and Stanley had three older brothers already in the business, so he was determined to show his worth by working while he went to school. Stanley was warm, lots of fun, well-liked and generous. He was always rushing to get somewhere on time, and he never saw a buffet he didn't like.

That summer, I also started to date this girl who lived in Bayside that I'd never dared to call when I was in high school. Unfortunately, on our second date, I made the mistake of bringing her home. I thought she was really cute and sexy, but I never heard the end of it from my mother. My parents didn't like her. I have no idea why, but they just didn't think she was right for me. I saw her a couple of times after that, but didn't bring her around home again. It was the first and last time I ever brought a girl home to meet my folks, until I brought home my future wife.

After my Buick had blown up in front of Sarah Lawrence, my father sent me back to Miami with a new 1954 Chevy Bel Air (a classic today). In our sophomore year, Bruce and I became even more inseparable. Not only did we room together and play all the ballgames together, but we now took the same classes.

I had been in engineering school in my freshman year. In the second semester I took a calculus class with a professor who was really a genius, one of the major scientists on the atomic bomb. Like a caricature, this professor would come running into class late, dropping papers behind him. He was difficult to follow in class because, being a math genius, he was just too far ahead of us. I found that while I wasn't able to work out the formulas to the problems the way they wanted, most times I came up with the right answers by doing different forms of arithmetic. Somehow I made it through this class, but my methods were far from satisfactory for the final exam.

When I had so many of the correct answers on the final without having shown the proper formulas, they suspected I was cheating, and made me take another test. I figured out the problems again in the same way and scored over 90%. They passed me, but made me drop the subject because I obviously hadn't grasped it. I, too, thought I couldn't grasp calculus, but if you look back on this today, you might say I was gifted to have come up with the answers the way I did. That was pretty unusual. The rules were more rigid then, and creative methods were not encouraged.

Without calculus I was forced to leave engineering school. While I wanted to stay in engineering, I wasn't shattered by having to change my major. I had already taken some business courses and enjoyed them immensely. Business was my second major.

My accident-prone streak continued into my sophomore year. One afternoon as I played softball, it was a tied game at the end of the ninth inning. There was one out, a man on third, and I was on second. A slow ground ball was hit to the second base side of the shortstop. I ran to third on the pitch, so when the baseman picked up the ball, he threw it to first. Out! I ran home, slid dramatically into the plate, and knocked the ball out of the catcher's hand. I was safe. We won the big game, but I'd skinned not just my knee but my thigh, all the way up to my hipbone on my right leg.

There was a doctor near our fraternity house, so the guys took me there. As we sat in the waiting room, I noticed that all the other people sitting there were very heavy women. I'd signed in like everyone else, but finally, when I was the last one left, the nurse asked, "Who are you waiting for?"

"I'm waiting for the doctor. I came to show him my leg." I pointed down at my ripped up skin.

"Oh, dear, this is a gynecologist's office," she said, trying not to smile.

We had been sitting there for two hours for nothing. I eventually did get doctored up, with no notable scars.

That same year, my knee took another beating and this time I wasn't so lucky. We were playing a football game against a rival fraternity who had a couple of ringers. (If you played on a college varsity team, you were not supposed to play intramural, like these ringers did.) They were big; one was at least 265 pounds. I was a 142-pound blocking back and defensive end. Bruce was a wingback and running back, so at 142 pounds, I was blocking for Bruce, who weighed 195 pounds at that time. This big guy, pretty much an animal, kept coming in. He seemed to relish demolishing me even more than going after Bruce, the runner.

On one play I was leading the blocking for Bruce, so I was running in front. This animal charged on the side and clipped me (hit me on the side of my knee). I could hardly walk, but I refused to leave the game because I was furious about the illegal hit. Bruce had seen it, too. In the huddle before the next play, he said, "This time let him come straight through." Instead of the ball being hiked to Bruce, they hiked it to the other back. I let the animal come straight through, and Bruce hit him from the side. Even though this guy was 265 pounds, they carried him off the field.

I ended up in the infirmary for a week, but they didn't do much for my knee. I returned to playing sports, albeit wearing a knee brace, but for the rest of my life I had a bum knee. Today, I have trouble walking any distance.

Finally, twenty years later, a friend recommended that I see an orthopedic surgeon who coincidentally was the doctor for a hockey team. I went to see this doctor and he X-rayed my knee; it was the first knee X-ray I'd had in twenty years. "I've got good news and bad news for you," said the surgeon. "The good news is that you were able to walk in here. The bad news is that you have bone on bone, no ligament and no cartilage. Because no one surgically stuck anything in there, Mother

Nature took care of it and grew calcium deposits on each side of the knee to hold it in place."

At that time, knee surgeries weren't normally as successful as they are today. The doctor recommended that I leave it alone for as long as it held up. He said that someday the calcium might break and that would be the time to have surgery, but the longer I waited the better it would be. I am still on that knee today, at the age of 69 when this book is being written.

Later that same sophomore year, I was in charge of intramurals for our fraternity. Every time you entered somebody and they showed up to compete, you got points. So, as unlikely as it seemed, I felt we could challenge for the President's Cup, the major sports prize in the school. I was amazed at how well we were doing, considering we didn't enter the full number of competitors for about half the events.

I rallied everybody in the fraternity to volunteer to play or do *something* to gain points. One of the sports was boxing, the hardest sport to get anyone to volunteer for. To set an example, I entered myself in boxing. Because I'd fought a lot as a kid I thought nothing of it. This was how I got my nickname—Rocky—and it stuck with me through the rest of my college career.

We wore helmets and big boxing gloves. When I got in the ring I was confident that I could beat anyone my own weight. The first fight was easy. The other kid lay down and I won the fight. My second fight was with a kid who really knew how to box, and he whipped the living daylights out of me. It made the newspapers and the headline read: "Klastapopowitz Slaughters Opponent."

Even though I was being pummeled in my second fight, I wouldn't go down. I was just standing up, and taking a beating. Like my mother, I was stubborn. I refused to give my opponent the satisfaction. Every time I went to punch him, he hit me harder in the face, so I just stood there, covering up. Other fighters would go right to the canvas when there was a mismatch, just get their points and get out. Because of the helmets and gloves I didn't get hurt too badly. However, after that experience, I didn't box anymore.

"In any competition the most important thing is not winning but knowing you did your best."—John Wooton, legendary UCLA coach.

Once I started heading up the intramurals for our fraternity and everyone saw that I was crazy enough to put myself into boxing, I got volunteers for everything. Our intramural points really started to rise.

My friend, Stanley, was good at all racket sports, so I entered him in the Ping-Pong competition. Stanley was about five-five with spindly legs, a southpaw (left-handed) with an awkward, unorthodox form. He appeared an unlikely opponent until you stepped on the court with him. Surprisingly quick, he was deceptive and never gave up on anything; he would find a way to beat you. Stanley wrote the book on sportsmanship.

In the Ping-Pong competition, our rival fraternity, Alpha Epsilon Pi, more than twice our size, entered a ringer—a junior in the school who was really a professional Ping-Pong player. Bob Alden put on exhibitions between halves in the Harlem Globetrotter Games in Madison Square Garden, where he did all sorts of trick stuff like hitting Ping-Pong balls into whiskey glasses.

No one had ever won a game from him. I knew when I set up the brackets that I had to keep Stanley from playing Bob as long as possible, keep him winning until the finals. There was no doubt that everyone in school would expect Bob to win, even though he was pretty obnoxious. When he played in the student center all day, he took anyone's challenges, including Stanley's, always winning.

Finally, we were coming up to the big event, Stanley against the ringer. At the "snake pit" where we all met between classes, the AE-Pi's were taunting us. "Why don't you bet?" they said. We wouldn't bet because we thought Bob would win easily.

One day, as Bruce and I walked across campus with Stanley, we were confronted by four AE-Pi's. "Hey, we'll give you five-to-one that he doesn't take even one game," they said. No matter the odds, Bruce and I still didn't want to take the bet, but since Stanley was with us, we wanted to show our confidence in him. We reluctantly put out $10 each, a lot of money back then.

The next day, the match was held in the large student center. A referee sat on an umpire's chair. Most of our fraternity was present, along with about 60% of the AE-Pi fraternity and some other players who were real Ping-Pong fans. The winner of the match seemed a foregone conclusion. Still, because of his arrogance, Bob was not well liked.

The match started—best three out of five sets. Stanley was playing about 40 feet back from the table and his opponent was smashing everything. Stanley just held up his racket and the ball kept coming back as if Bob was hitting it against a back-wall. The first game was even all the way. Finally, Bob started to get frustrated. They got to the final point. Stanley had frustrated his opponent so much that Bob momentarily lost his cool and overshot. Stanley won the game. We went crazy, cheering.

We got our money, but that was the least of our victory. This event gave us major points for the President's Cup. AE-Pi, our arch rivals, were ahead of us going into this match. Our expectations grew. If we won the Ping-Pong championship instead of coming in second, we'd have a great chance to close the lead.

Stanley lost the next set, but won the third. By this time, word had spread around campus and the student center was filling fast. The excitement in the room rose to a fever pitch. They'd each won two sets going into the fifth. By the last game of the final set, the score was even. This was it, the last volley; we were all holding our breath.

Bob smashed the ball off the table. Through the cheering, Bob was screaming that the ball had ticked the table. The umpire made the call in Stanley's favor. We had won! Typical of Stanley, he put up his hand to quiet the crowd. "Listen, I don't want to win that way. Let's play it again."

Play It Again? When Stanley came over to the sideline, we almost pummeled him to death. Everyone was mad and saying things like, "What the hell are you doing?" and "What are you thinking?" We were shaking him.

"Never give a sucker an even break," Bruce said.

"Don't worry," Stanley said calmly. "I'm going to win."

Sure enough, he won the next point and the set. The Ping-Pong championship catapulted us towards our run for the President's Cup.

That Ping-Pong match was one of the most exciting sporting events I have ever seen. It sounds funny to say that about college Ping-Pong, but given the personalities of the two players, it really was that exciting.

Remembering Stanley's declaration that day reminded me of another famous quote:

"You Got To Believe."—Tug McGraw, 1969 Miracle Mets.

Another individual sport in the intramurals that year was wrestling. Bruce had gone to Mepham High School in Merrick, a major school for wrestling. He hadn't wrestled competitively on the team, but the school had taught the sport in their gym program. I used the same strategy with Bruce and wrestling as I had with Stanley and Ping-Pong.

Pratt, a 270 pound, all-American football player was wrestling for his fraternity. Every match he had was over in only one minute; he just picked people up and slammed them down. The interesting part is that Bruce wasn't supposed to wrestle heavyweight. Up to 195 pounds was classed as light-heavyweight. I stood in front of Bruce when he weighed in and read the scale. It actually said 196, which would have put him over, but I was going to say 194.

At that moment, the director of athletics walked up to pose for a picture along with Bruce as the best intramural athlete at that time. I couldn't help him. The scales read 196, so Bruce was put in the heavyweight division to fight these football players and other huge guys.

Match after grueling match, by sheer willpower and training, Bruce made it to the finals. There he finally met the monster, Pratt. Pratt picked Bruce up and threw him around. Bruce wouldn't let himself get pinned. If no one gets pinned, it goes on points.

Bruce was way down on points and the match was almost over. Suddenly, taking his opponent by surprise, Bruce turned around and pulled Pratt down with a thud. With every ounce of strength and willpower he had, he pinned that big, huge football player. After the referee counted his three, Bruce lay on top of this guy, limp with exhaustion. He won! Then, like a bench press, Pratt picked him and angrily tossed him across the ring like a rag doll. It took four or five of us to carry Bruce out. We put him in the car and took him back to the house, where I sat him on a stool in a cold shower. He sat there for well over an hour before he could even pull himself up. He was exhausted.

After our wrestling victories, we had to win two more team sports for the President's Cup. Our best chance was football.

We played the last game of the season against a big rival, the other undefeated team. Their quarterback, Phil Morgan, had been a star quarterback for a Connecticut high school. Coincidentally, Bruce had just begun to date Peggy, who had previously dated—of all people—Phil. Bruce would eventually marry Peggy, but he had no love for Phil Morgan on the day of our big game.

I was playing defensive end. Their quarterback ran the whole game on option: when you would go after him he would toss the ball off, toss it over your head, or run with it. It was a tough game. With a minute left to play—the score was seven-six for the other team—they had the ball on their own five-yard line. They could run out the time on the clock and win. Bruce came to me and said, "Phil Morgan has such an ego, he is not going to stand still and just be tackled. Go straight for him. If you can chase him into the end zone, I'll take care of the outside."

Sure enough, I cut in and Phil started to run with the ball. He ran wide, into the end zone. I dove and caught his heels with my fingertips. We got a safety (two points), thus winning the game by one point.

"It's not over till the fat lady sings." — Yogi Berra.

We won the football championship! That was my moment of sports glory. To top it all off, we won softball also. That year we won the President's Cup. It was unheard of for a small fraternity like ours.

Athletic competition, both as a boy and in college, has served me well in business, as well as in life. College was basically the end of my athletic endeavors for almost twenty years while I earned a living and brought up my family. I was nearly forty before I took up tennis, and shortly after I began playing tennis, I also returned to golf.

CHAPTER 7

Co-Eds

Our third fraternity house roommate, Howie Strauss, was also a New Yorker. The summer between our sophomore and junior years the three of us continued our friendships back home. Howie was a very wealthy boy. His father was the largest prophylactic manufacturer in the country. When we had our fraternity parties on Fridays, we had all these little white balloons hanging all over the house.

Howie was tall, good-looking and fun. He had a prep school manner that made it evident he had been born with the proverbial silver spoon in his mouth. Unfortunately for Howie, he lacked the backbone to stand up for himself, particularly where his "our crowd" father was concerned.

That summer, my mother met Howie and liked him. I finally succumbed to her matchmaking pressure by introducing him to my cousin, Ann, and they became an item. To return the favor, later in the summer, Ann fixed me up with a pretty girl she knew from Sarah Lawrence. Howie invited us all to his parents' huge apartment—the entire floor of an elegant building on Park Avenue and 58th Street—for a drink.

Of course, Aunt Ruth heard of the upcoming event and gave my mother the usual warnings about me. "Make sure he doesn't do anything crazy. Tell him not to embarrass Ann. She got him this date."

That night, I drove into the city with the two girls and parked on 58th Street, just off Park Avenue. I didn't bother to lock my car. In the 1950s, as it still is today, this Park Avenue neighborhood was about as fancy as it gets in New York City.

When I returned to the car with the girls following a few steps behind me, I started toward the back door, preparing to open it for them. I stopped cold. There was a nude man lying on the back seat of my car, and it looked like he was asleep.

Nervously, I wondered if I should open the door and holler at him to get out, or what? "Stay back," I told the girls. "You're not going to believe what's in this car!"

By this time, Howie arrived and looked in to see the naked man. He was furious. This kind of thing didn't happen on *his* block. He swung open the back door and screamed, "Get out! Get out!"

The nude guy blinked, then reached under the front seat and pulled out a knotted bundle of clothes. He put on a sailor hat, then climbed out and staggered down the block, stark naked, with his clothes under his arm. The girls stared in disbelief.

My parents were up when I got home, and they asked, "How did your date go with Ann and her friend?"

"Okay, I guess. It was kind of weird. There was a naked man in the car." By this time my mother had heard so many outrageous stories—some of them exaggerated, some not—so I used to downplay things. Early the next morning, the phone rang, and of course, it was my aunt berating my mother about my behavior. After my mother hung up the phone, she said, "Why didn't you tell me?" I did," I said.

"Well, I didn't believe it," my mother said with a sigh. Her life was never dull with me in it.

Howie didn't return to school with us the next year because his father got him a cushy job in the Navy in order to avoid the draft. Bruce and I returned to Miami for our junior year, Stanley for his senior year.

Bruce and I were in all the same classes, and now had been rooming together for almost three years. The year before, a coed had started a rumor that we were gay because we were together so much of the time. Being gay was such an odd thing to us back then that it wouldn't have entered our girl-chasing minds in a million years. We thought the idea was funny and treated the rumor like a big joke. Every once in a while, when we'd pass a couple of girls in that sorority, we'd hold hands.

When we went home for Christmas that year, Stanley and I got in touch with Howie Strauss and made plans to go to the Copacabana to see Frank Sinatra. It was a big deal and we were bringing dates. I'd had a couple of dates that semester with this Florida girl who seemed to like me, but I didn't have a date up north. My cousin, Ann, was still seeing Howie.

I didn't want to be embarrassed, and I was determined to get my own date, rather than have my cousin fix me up. I called my Florida girlfriend and she said, "Oh, I have a friend who's in Brooklyn now. I'll fix you up with her."

"That sounds good," I told her. "Stanley mentioned his date lived in Brooklyn too."

Several nights later, I met Stanley and his date in Brooklyn, then we went together in one car to find where this other girl lived. It was in a six story walk-up, and she was on the top floor. "If she's a winner, I'll give you a thumb's up," I told Stanley who walked behind me as we climbed the stairs, "and if she is a loser, I'll put my thumb down."

I knocked on the door. My mouth dropped when it opened, and not in a good way. My date was about five by five—literally—she looked like one of these old-fashioned standing TV sets. To make matters worse, she smelled like a plucked chicken. She had just tried to light her gas oven and had burned the top of her hair and her eyebrows off!

Stanley had warned me against the wisdom of letting the girl I was going with fix me up with another girl. I should have listened. I barely knew what to say. "Uh, maybe you shouldn't go because you had this accident. Forget about it. We'll . . . do it some other time," I managed to stammer.

Meanwhile I didn't have the presence of mind to remember our signals, so my thumb was still up. From behind me, Stanley couldn't see the girl. "Just take her. Just take her," he urged, thinking she was a looker.

As I stepped away from the door, Stanley peered in and was in shock. I didn't know how to tell her *not* to come. The girl tagged along, but I was dreading the evening. As my cousin Ann would be with us, I knew it would get back to my mother and my aunt and I'd never hear the end of it.

Before the show we stopped by Howie's Park Avenue apartment on 58th Street. On the way, I was feeling more and more remorse at not being able to get rid of her. At Howie's place we hung up our coats in a closet that was almost the size of a room. (An odd thing for a closet, but this one locked from the outside.) Suddenly I couldn't help myself. When my date walked into the closet to hang up her coat I locked the door behind her.

As I walked into the living room, Howie asked if I wanted a drink. Even though I rarely drink and never drink whiskey, I silently took the large glass he handed me. He dropped in two cubes of ice, and began pouring. "Say when."

I didn't say "when." I just stared at the amber liquid growing higher and higher in the glass. "Are you really going to drink this, Don?" he asked.

I drank the 12-ounce glass of rye, barely aware of the burn going down. I was so stunned, and in such shock that I didn't even feel dizzy. The only way to get out of the situation, I thought, was to pass out. *Please, just let me pass out.*

Just then, we heard banging on the closet door. "What's that?" asked Howie.

"My date," I said, surprisingly articulate. "She must have gotten locked in the closet. I'll have to let her out." I didn't get up.

Stanley was looking at me strangely; *he knew*. Howie got up and let my date out of the closet. More than ever I wanted to pass out and disappear, but we all went on to the Copa as planned. We tipped the maitre d' to get a great table, and it had a big column in the middle, which I sat my date behind. That evening was one of the most embarrassing events in my life.

Back at school after the Christmas break, Stanley, Bruce and I were forging ahead so well with all the business courses we needed in order to complete our majors that the three of us decided to take typing. We figured if we knew how to type, we wouldn't have to pay to have our papers typed any more.

Typing was taught two nights a week, Tuesday and Thursday. After class, we just had time to catch the jai alai games, a sport we also used to play. The only trouble with the typing class was that I had about ten thumbs. The typing teacher finally told me that I was undoubtedly going to fail. I had never failed a course before.

Stanley was doing especially well and had become friendly with our teacher. "Let me talk to him," he said. I was then given a second chance to take the test, but I still couldn't type enough words in the time allotted to pass.

The typing professor was definitely gay, no ifs, ands or buts about it . . . and he liked Stanley. In order to get me a passing grade, Stanley took the teacher to dinner and to jai alai, and I got a passing grade under the condition that I drop the course and never take typing again. Stanley made the ultimate sacrifice; that's what friends are for.

One of the most eventful things that happened in my junior year was a trip I took to Cuba. It didn't cost a lot of money to fly from Miami to Cuba, and at that time Cuba was an open country.

Usually there was a big winner or two in the all-night poker games and crap games at the fraternity house. If there were two or three winners they would often pool their winnings and take a flight to Cuba together. I went with my friend, Artie Kroll, but he never gambled. He saved every penny, and when he'd saved up enough for the ticket, we went to Cuba together.

We stayed in a big room at the Hotel Rex with high ceilings and a shower right in the middle of the room. It had one other tiny room with a toilet and what Artie thought was a drinking fountain in it. I knew it was a bidet because I had traveled with my parents and seen them before. Artie was very short—about five-two. He kept drinking out of the bidet, and I let him do it the whole trip.

The main reason for going to Cuba was to visit the local houses of prostitution, which were totally legal there; they even had doctors to see that everyone was healthy. Artie and I were all excited about visiting these houses, having heard about them from our fraternity brothers who had already been to Cuba.

Just my luck, when we arrived the Seventh Fleet had landed in Havana, so the houses weren't as stocked up as usual. The first night we got there, we were too late—I guess most of the girls had been spoken for—so the next day we went early in the morning. The girls all parade in front of you, quite a show in itself. They flirt with you because they want you to pick them so they can work that night and earn money. Artie and I finally picked two girls. We told them we weren't going to go with them right then because we wanted to go out to eat and spend the day exploring the city. Paying money to reserve the girls for later, we made plans to pick them up that evening.

They were all dressed up and smelling of strong perfume when we returned to pick them up and take them back to our hotel late that afternoon. I was never big on perfume, but it wasn't until some years later that I realized how allergic I was to it. At the time, I just thought I didn't like the smell of perfume.

Artie and I kept these girls with us until two o'clock in the morning. By that time we'd both had enough and just wanted to sleep, but the girls wouldn't leave us alone. Finally, we asked them to go home. "No, no, no. We don't go home," they said. We finally had to pay them not only their cab fare but more money just to get them to leave; they got us coming and going.

The next day we decided we'd had enough of that and would go sightseeing and buy some gifts. Cuba was a duty free port so you could buy liquor and perfume very inexpensively. We went into a fancy shop. They didn't speak much English, but I asked about the perfumes. "What's the best?"

The clerk said, "Chanel No. 5. Here, smell it." As he squirted me with this perfume I realized it was the odor I'd smelled the night before. I let out a holler. It had taken us half the night to rid ourselves of it. Artie and I had scrubbed ourselves in the crazy shower in the middle of the room. Now, this clerk had just squirted me with it!

He must have thought I liked the scent, so he started squeezing the atomizer again. I ran out of the store and I don't know why but he chased me down the street, squirting me with perfume.

At the end of our junior year, I had made the Dean's list and finished my major. Bruce's parents and my parents decided to give us our graduation gift a year early—they were going to send us on a trip to Europe. Hearing this news, I was very excited, but Bruce wasn't keen on going to Europe. Every year, from the time he was old enough, he'd gone to camp and he loved being a camper.

I never liked camp because I couldn't stand the regimentation, or having only fifty minutes allotted for ballgames. For the past two summers Bruce had worked as a counselor. He told me that if I'd go to camp with him I'd also get a job as a counselor. I must have been demented, but somehow I allowed him to talk me out of going to Europe that summer, and into going to camp. When I told my parents they were surprised, but happy to let me go.

Camp Equinock was known for being a very athletic camp. Bruce had heard they needed an archery instructor, so in order to be hired as a counselor I had to tell them I knew archery. I really did have some archery experience, because in Beechhurst my father had set up a target on the lot next door. I wasn't good at archery then, and now I had to teach it. I decided to buy a book and read up on it.

The summer of 1954, off I went to Camp Equinock. A competitive, athletic camp, it was built on the lakeshore in Pennsylvania. I was a senior counselor in a cabin of 10 or 12 nine-year-old boys, assisted by a junior counselor.

I have to say I was a very good archery teacher. However, I never took a shot in front of the kids because most of them were better than me. I couldn't hit the target half the time, much less a bull's eye, but I was able to teach them the technique of how to do it without actually demonstrating. I developed this one trick shot behind my back, which hit the target half the time. Even if I missed, it was no big deal because everyone was impressed by my amazing trick shot.

Let's face it, archery was not a very popular sport, but all these kids loved being with me. They would come up to the field during free play—20-30 kids every night, even on my days off. I would stay in the camp with the kids on my days off because there was nothing else to do and I really enjoyed these nine-year-old boys. All the kids loved me. I enjoyed teaching them and listening to them, and I guess they could see that.

One evening, near the end of the summer, not one kid showed up at the archery field for free play. I waited around, but no one came. I stayed a little longer, then I started to shoot at the target myself. I hit maybe one out of five shots. All of a sudden, I heard roaring laughter in the bushes, and about thirty or forty kids were rolling around hysterically on the ground. They'd caught me!

As I said, I was very involved with the kids—one especially. His name was Harold. He was wiry and athletic, but painfully shy and spoke with a slight accent. His parents had left Europe during, or right after the war. I taught Harold how to play ball, watching proudly as he gained more and more confidence. By the end of the summer, he had really come out of his shell. He became the best ballplayer in the camp.

The word was out among the counselors that if you were pinpointed as a really good counselor you could take home up to $1,000 worth of tips when the parents arrived on visiting day. That summer, I went home with $2,800. In the mid-fifties that was a huge amount of money. Harold's parents alone gave me $500. I made more money at camp that summer than I would have made working at Rexart.

When I returned for my last year of college I had already finished my major and had all the credits I needed. My mother, who'd always felt I lacked culture begged and pleaded with me to take some cultural courses. I gave in and signed up for music appreciation, drama, English literature—subjects I would never have chosen on my own.

At the beginning of my senior year, I became a senior adviser for the freshman class, helping them make out their programs. Since Bruce and I were leaving this school, we decided to get our fraternity a really promising rush class and ensure that Phi Sig would remain a strong group. We became pledge masters in charge of the new freshmen rushing our fraternity. Like shepherds, we would first help them get into the fraternity, then guide them through their first year.

I was no "big man on campus," but I was a senior in charge of rushing which made me someone to look up to. One evening, five freshmen at a rush party asked me where to meet girls. "We'll go to the student center," I told them, although I rarely did this.

I led the way, and just as we reached the student center, six freshmen girls were entering the building. I introduced my charges and got the girls talking. We were six guys—me and five freshmen—and they were six freshman co-eds, so I ended up talking to one of the girls, Judy Gottlieb, who had all the eye-appeal a college boy would want. Both pretty and cute, she had brown eyes and short brown hair, a great smile, and wore a white blouse and straight black skirt. My first impression was that she seemed more sophisticated than the others. Best of all, she laughed at my jokes, even though I wasn't sure she understood them.

Judy, who would later become my wife, had only been at school for one day then. I explained that I was a senior adviser who could help her with her program. The next day, she made a point of coming to me to sign up for her classes. Since I was no longer taking advanced business courses, I put her in all my first year cultural classes, even though her major was Education. From then on, since we took the same classes, Judy and I saw each other every day.

We started dating almost immediately. Judy was always very reserved, refined, seemed to have a lot of common sense and was fun to be with. And she was cute. Judy became my Friday night football date, and I took her to the fraternity party afterwards. She always told me what a terrific football fan she was, and raved about the games. It wasn't until after we were married that I learned she didn't even know what a goal post was. She'd accompanied me to the games every week, and simply cheered when the rest of the crowd did.

Bruce and I had moved out of the fraternity house the previous year and now lived in an apartment in Coral Gables. The two of us roomed with another frat brother, Marvin Epstein. Stanley roomed with Artie Kroll (my Cuba buddy) and Sy Levin, nicknamed Tojo. We lived around the corner from each other.

It was a twenty-minute car ride to the college from our apartment. I was always on time for the eight o'clock music appreciation class. Judy, who lived in a dorm just across campus, would consistently be late. When I glanced out the classroom window, I would often see her running across the grass with books tucked precariously under her arms.

Our professor started off the morning by playing a very calm piece by Mozart, Beethoven, or Chopin, and the whole class would sit there with closed eyes, relaxing. The seats and desks in this classroom were made of aluminum and there was a little catch on the desk to fold it down. Every morning, as the rest of the class listened to relaxing music with their eyes closed, Judy tried to sneak in quietly and put her books on her desk. Every morning I hit the catch on her desk and the books slammed to the floor. Everyone jumped up, startled. Judy was always flustered and embarrassed. This went on for several weeks until I stopped. She never figured out what I'd done (until she read this).

Whether she knew it or not, Judy put up with a lot from me at times. As I was graduating regardless and didn't need the credits, I didn't pay much attention in class or study for the music final. She did. During the final exam I started to write down the answers, then glanced over at Judy's paper to see if they were correct. I embellished her answers and ended up with an "A" while she got a "B." She was furious. All my life, I could always tell a good story.

When we first were courting, I would tell Judy outrageous stories. I treated her almost the way I treated my mother and loved to play jokes on her and pull her leg. She was very gullible, especially at first, so I had fun doing it. I finally told her so many crazy stories that she wouldn't believe anything I said. Just like my mother and the story of the naked man in my car, Judy wouldn't believe me when I was telling the truth. From then on, I always told her the truth, but even after we'd been married for many years, she still wouldn't know whether or not to believe me. She took my words with a grain of salt.

Early in our relationship, I told Judy of all the races I'd won with my sailboat, and naturally, got a little carried away in the telling. As I mentioned earlier, I was a sailor who raced Comets on the Long Island Sound, so I really did know how to sail. Boasting to a pretty, likable girl isn't unusual for a young man.

One weekend, about two months after I met Judy, I was going to take her and one of my frat brothers out sailing. At the boat rental place in Miami, they brought a catboat—a little boat with a sail—up to the dock for me. I was dressed in white slacks and a short-sleeved yellow shirt, looking very dapper.

"We don't need bathing suits?" Judy asked.

"No, you won't even get wet," I assured her.

I put one foot on the dock and the other foot on the boat. Suddenly, the boat drifted away from the dock. I slipped down between the boat and the dock, getting slimed green and totally wet. It wasn't cool. Red-faced, wet, and with my white slacks covered in green muck, I took Judy and my buddy sailing. I spent the rest of the afternoon attempting to prove I knew what I was doing.

Back at school, our drama class featured guest lecturers, and we had scheduled such playwrites as William Faulkner, Arthur Miller and Tennessee Williams. The most memorable was Tennessee Williams, probably because of his heavy Southern accent. He wore a white suit and was quite a character.

The drama lectures were given in a large lecture hall, set up theater-in-the-round style. One day, some law students came in with a petition saying that blacks—referred to them as "Negroes"—should be allowed in the school. They passed the petition around the class for signatures.

In 1956, the University of Miami was a segregated school. I had a lot of friends in the athletics department that I had gotten to know through my involvement in the school athletic program, as intramural director for our fraternity. All of them were Southerners, and some were former professional ballplayers. I knew how strongly they felt about keeping out integration.

When the petition came to me, I thought: *I'm leaving this school soon. Why should I have a say in this? Why should I impose my will on anyone else?*

I passed the petition on without signing it.

Judy had grown up in Manhattan, and was a counselor at an underprivileged camp mostly attended by black children. She was terribly embarrassed that I didn't sign the petition. In hushed tones she began to argue with me right there in class.

Almost everyone in the room had signed the petition. Three of my fraternity brothers, who were also taking the class, questioned me. Judy was still arguing with me as we funneled out of the room through the big crowd. Two of the law students stood at the door, and when I got there, they screamed at me, "Why didn't you sign the petition?" They made a big scene, shoving the petition in my face as they carried on.

My fraternity brothers laughed and Judy looked as embarrassed as hell. Having a low boiling point, my patience snapped and I grabbed the papers with all these names. I tore them into little pieces and threw them up over the law students' heads like confetti. "Happy New Year," I said, and strode from the room.

I went with Judy Gottlieb all that fall. Instead of my going home for Thanksgiving, my parents came down to Florida with Roger for several days of vacationing. At this point, they knew I was going with someone. They were staying at the Eden Roc Hotel, and they invited us to dinner. I hesitantly brought Judy to meet them, the image of the only other time in my life I'd ever introduced them to a girl still in my mind. Nonetheless, we went to dinner.

My mother was her usual charming, funny self with Judy. It was love at first sight, and I think my father felt the same way, although he was quieter. From that moment on, my mother started to push Judy on me, the exact opposite of the way she'd been with the other girl. She'd make sure I was treating Judy well. "What are you buying her for Christmas? You've got to buy her a gift," my mother would insist. Amazingly for me, I vaguely remember giving Judy the blue scoop cashmere sweater she lovingly remembers; my mother must have bought it.

After the dinner, I asked Judy for her first impression. She thought my mother was very cute. My parents were so all over her that night that it was obvious they liked her. Judy's mother came from Latvia and her father from Poland. My parents looked very young and very American to her, whereas her parents were much older. When Judy was born, her mother had been about 37 or 38, and in those days, that was late to have a baby. All of Judy's girlfriends had very young American parents.

The Gottliebs owned a variety store in Manhattan—a five-and-dime store, they used to call them. Judy's mother, Minnie, whose maiden name was Katzen, was a working woman who ran the store. Her father, Ruben, had worked for her mother, which was how they met and married. Judy's mother had sisters and brothers who were bright and had good business acumen. They were also in variety store businesses in Manhattan, and four had moved to Miami Beach to run successful small hotels in what today is fashionable South Beach.

That Christmas, when we got home, my parents invited Judy to the house. They were really giving her the rush job. My mother made a special dinner. I didn't recall this, but Judy told me later that my mother was carrying on the whole conversation that night, and my father hardly opened his mouth. This is very similar to the way I act when there's someone new or when I'm in a new situation.

Near the end of her freshman year, around March or April, I pinned Judy. That meant that I gave her my fraternity pin, which was like going steady, a sort of pre-engagement. I was graduating and leaving school that spring. After I pinned her, the graduating class had a formal in Key West. Several funny things happened there.

My fraternity brothers had gone down to Key West for a three-day weekend formal. Judy was voted "Sweetheart of the Fraternity" for the year, and my fraternity brothers serenaded her. Stanley's future wife, Iris, roomed with Judy, while Stanley and I roomed together. The only thing Stanley and I got out of it was that we accidentally switched one of our shoes. A number of years later, we finally realized we had each worn one of each other's shoes all that time.

On the morning of the formal, Stanley and I went down to breakfast to meet Judy and Iris, but they weren't in their room. We went looking for them, and finally found them outside by the pool, but they weren't sunbathing. Judy had lost my fraternity pin and they were frantically searching for it. Somehow, with all four of us looking, we found it on the ground. Judy was sure relieved.

At the formal party we gave out a party favor, a black stuffed poodle that was about fifteen inches high. That night, Judy wore a special dress with eyelet lace that she'd asked her mother to buy for her. For some reason, she carried around this black stuffed poodle on a red leash all night long. "Judy, put the dog down," I said as we danced. She didn't reply, and wouldn't put it down. I later found out it was stuck to her new dress; the leash had caught in the lace.

Over the years, this poodle wasn't the only thing that got stuck to Judy. She has a history of things sticking to her. But this was the precursor, a vision into the future.

CHAPTER 8

Real World

After I graduated from college that summer I went to work for Rexart. I had majored in Systems and Procedures, so basically I was an efficiency expert. The first thing my uncle did was tell me to make a twelve-page report on what I would do to improve the place. He gave me this assignment on a Friday, and wanted it completed on his desk Monday morning. I worked all weekend on my twelve page report. Feeling pleased with the job I'd done and eager to show my uncle, I brought the report in to his office bright and early Monday morning.

Uncle Mort was an engineer who had attended officer candidate school at Yale. A former officer in the Air Force who'd worked on the "buzz bomb" in World War II, he was a fair but tough boss. I got along terrifically with all the veterans he'd hired. I had played ball with them in the summer leagues and worked with most of them before as a kid. Now I returned to Rexart as an adult, as an efficiency expert with a business degree, looking forward to working with them all again.

Uncle Mort glanced at the cover, then without even opening it, he ripped the report in half and threw it in the garbage can. "This is the real world. Welcome to it, Donny."

I bit my tongue and managed not to say anything I might later regret. It was a shocking welcome to the real world in my first week on the job.

By this time Rexart was hooked up with the union, run by the notorious Harry Davidoff. A few months after I did my report the company had major union problems. The shop stewards were slowing production down by insisting no one work faster than the slowest person. The company needed more production.

Rexart had one very hot item that year, a chaise lounge, but they only had one location to make it and only so many hours to work. The union wouldn't allow for overtime or anything else to increase production. Treading carefully, I suggested that maybe a time-and-motion study was in order. "They're assembling about 122 lounges a day now. Time the production and give them a certain amount more to produce," I told Uncle Mort and Dad.

My uncle was a "doubting Thomas" when it came to new ideas, but he said, "You think it's a good idea? Okay. I'll bring in Davidoff and you talk to him about it."

Davidoff stole just for the fun of it. If they left Davidoff alone in their offices, he would break into their desk drawers and steal pencils, paper clips, anything. I was already seated in the tiny room when he was escorted in, then left alone with me.

"I got an idea how the people in the chaise department could make some more money," I proposed, then began to explain how we would time their production with a stopwatch, and . . . .

As I spoke, I watched Davidoff's face turn redder and redder. He had these hands like little fat sledgehammers, with a chunky gold diamond ring on every finger. I wondered if they were trophies from his victims.

He listened quietly until I was done, then said gruffly, "I got a better idea, boy. Suppose you take out a stopwatch, and I put you in concrete boots and drop you in the middle of the Sound?" *This was the real world.*

Knowing his reputation, I understood that he wasn't fooling. Thinking fast on my feet, I said, "Listen, just to simplify the whole idea, and show you it's going to help your workers, not hurt them . . . suppose, for every extra chaise lounge they pack over 122, we give them a dollar more."

"Yeah, and then what do you do?" he challenged.

"Nothing," I said. "The deal is: the production group gets paid a dollar more for every chaise lounge they make over 122 per day."

Davidoff wasn't sure he even wanted to accept that. He reluctantly agreed, with a warning that, coming from him, sounded like a threat. "They better not lose money; they better only make more money."

"They can't lose money," I assured him. "And if their production goes up they *will* make more." Frankly, I don't think he understood too much.

That was my first encounter with time and motion studies in a hands-on business situation. I put the study into play in the department where they had 28 people putting together and packing the chaises. Their production rose gradually from about 122 to 170-174. *Every day.* It was a tremendous increase, but I'm sure my uncle thought it was an accident that my idea worked.

However, I discovered that I couldn't get too complex or technical. Keeping it to its simplest common denominator enabled everyone to understand how it worked. Many of the fellows working in the department didn't even speak English, but it was easy to explain to them that for every chaise over 122 that they produced, they would get an extra dollar to share amongst all of them. That extra money would end up in their pay at the end of the week. Production kept going up until it hit the mid-170s a day. We were happy with the increased production and the workers were happy with the extra money.

That was my first big success in that field, but eventually my profession would become obsolete. One of the stories about why efficiency experts became obsolete shows how I almost became obsolete, but first let me tell you more about Judy.

That summer, I asked Judy not to go back to the university and she agreed. She hadn't done fabulously in school, although she did okay. Her mother had sent her to the university to get married anyway, so she supported this decision. Judy stayed in New York and went to secretarial school, which eventually led to a good job with a publishing company. Unlike me, she typed very fast.

Judy lived on West 95th Street, and I lived in Beechhurst, a 50-minute ride apart. After our dates, I had to drive all the way back to Beechhurst. One night, I fell asleep at the wheel of my Chevy Bel Air. I must have been sound asleep because I crossed First Avenue. A huge milk tanker truck hit me, demolishing the car and flinging it up onto a concrete island. This was before the days of seat belts, but I was so soundly asleep that I was totally relaxed. The only injury I sustained was a slightly bruised leg.

After the accident, my brother, Roger, drove with me to pick up Judy. She lived in an apartment house, across the street from a row of old brownstones where a lot of Spanish-speaking people lived. At this time, many Puerto Ricans were moving into New York City, and the neighborhood where Judy and her parents had lived for many years was becoming a Spanish area. Judy's parents owned a store around the corner from their apartment, so they didn't want to leave because their business was here. It was fast becoming a scary neighborhood with all these tough-looking Spanish guys sitting on the stoops of the brownstones.

Judy and I kept dating and things were getting more serious. Before Christmas, my mother asked me if I was going to give her a ring. "She's going to expect it," she said.

"No, I'm not giving her a ring now. I'm not getting married yet," I said.

"Well, you'd better give her a good present," my mother said, and suggested pearls as an appropriate gift. With my parents' help, I bought Judy a string of pearls. Over the holidays, I took her out to a fine restaurant. I ordered oysters, which were served under a silver cover, and had them put the pearls in with the oysters. Judy was very excited to find them.

By then, both Bruce and Peggy Tucker, and Stanley and Iris Rabinowitz had gotten engaged. Bruce and Peggy had even set a wedding date. We had all been going to the university together, and had dated in Miami and at home; all six of us were close friends. But that same Christmas, my fraternity brothers and I had the draft breathing down our necks.

Since we would have to do our two years of military service, I suggested to Bruce that we ought to join the Navy, like Howie had, before we got drafted and ended up not getting a choice. The Navy had an officer candidate program where you went in for four years, and they allowed you to serve on the buddy system. Bruce and I could be on the same ship.

Just before Christmas, we went down to the Navy recruitment offices and talked to the recruiter. Afterwards, we discussed our options and decided we were going to sign up.

Two days later, I went back to the Navy recruitment office in the afternoon when Bruce was supposed to meet me. The recruiter we'd talked to was off Christmas shopping, so I didn't get to sign up that day, and for some reason Bruce failed to show up.

I called Bruce to find out what happened, but couldn't reach him. He called me the following day to tell me he'd been drafted. Once you were drafted by the Army, you couldn't join any other branch of the service. It was too late.

Meanwhile, I had very nearly joined the Navy, and Bruce wouldn't even have been there with me. If that recruiter hadn't gone Christmas shopping, I would have been hooked for four years. I decided my best option was to wait for my draft notice and go in for my required two years.

Bruce was so nervous that he broke out in a rash which delayed his physical for about four months. By that time, he and Peggy had gotten married, and she was pregnant, which allowed him to avoid the draft.

Stanley was declared 4F and married Iris shortly after. I was waiting to see what happened with the service before I proposed to Judy.

I had a subscription to *U.S. News and Business Report*. The February issue announced a new program, that was to start in April, where you could enlist in the Reserves for six years and go into the Army for only six months. The minute I read it I thought it was a good idea. I showed it to my father and talked with Judy about it, and they both agreed. The next day, I went to Fort Totten, which was right in Bayside and had a Reserve unit then. I signed up for the Reserves.

I now knew how long I would be in the Army and when I would be able to come home and start a life. I proposed to

Judy on Valentines Day, which was also her birthday. I took her to a nice restaurant on Central Park South called Pen and Pencil. There I had a silver domed serving dish delivered to the table with a 2.25 karat heart-shaped diamond ring under it. Before Judy even said *yes*, I got my answer in her amazing smile and the way her eyes lit up. After dinner, we took a romantic carriage ride through the park. We ended up near Judy's house and walked there to tell her parents that we were engaged. They were thrilled.

Her mother's cousin, a jeweler, came over to her parents' apartment. From where I was sitting in the living room with her father, it just happened that I could see the big mirror in the hallway, which reflected into the bedroom. In the mirror, I saw her mother's cousin examining the quality of Judy's engagement ring through a loupe.

I was so nervous and stressed out from that scene that I went home and couldn't even get out of bed and see Judy the next day. It took me a whole day to pull myself together; Judy finally came over to visit me.

Iris and Stanley's wedding was set for March 17th. When I signed up for the reserves I thought I would be going into the Army on April 1st. However, since I'd signed up so early, our group left before the whole program officially started. Judy attended the wedding without me because, that very day, I was on a bus on my way to Fort Dix.

I ended up in a very unusual unit because of the special program. About 108 of the120 men in my company were college graduates. They all had read about the Reserve program and jumped at the opportunity to enlist right away for only six months rather than two years, even though we all would have to go away for two weeks every summer for the next six years.

In boot camp they worked our butts off because the noncommissioned officers (NCOs) were upset that we only had six months of active duty. They ran us ragged and we were dropping like flies. With all the running with heavy backpacks for miles, rigorous calisthenics, and general training overkill, we weren't breaking any records. When they finally gave us something we had to study, a group of us came up with the idea to make a deal with the captain: We'd give them the highest scores they'd ever had on the written exam if they'd lighten up on the physical stuff. They agreed, and we gave them their record-setting scores.

It was too late for me. I was so run down that I got pneumonia. When I went to the infirmary they turned me away, didn't even give me antibiotics. About three days later, I was in terrible shape. It was a Sunday, my mother was coming up to visit, and I didn't have the strength to get off my cot to get dressed. But you didn't want to lay around your barracks in the Army, especially in boot camp, because someone would come around and volunteer you for something. Sure enough, I heard someone coming. I pulled myself out of bed and hid in a tiny, narrow locker. I eventually got dressed and waited on the steps to meet my parents.

Taking one look at me, my mother sent my father into town for a thermometer. I was running a temperature of about 106. "Where's your sergeant?" my mother demanded. I pointed the way, and she went directly to the sergeant.

"How in the world . . ." she began her tirade, taking a breath only when she'd finished. She wiped the floor with the first sergeant.

I was put in the hospital where I stayed for two weeks, the time it took for me to recover. Judy and my mother came up to visit often. Everyone thought my mother was my mother-in-law and Judy was her daughter, even for years after, because they had a similar look.

After two weeks, they released me from the Army hospital and "recycled" me. I was put back to the beginning of the boot camp training with all the marching. I was informed that I would never be able to go back to my original group because they were now far ahead of me. This time I was put in with a group from Maine: big strapping farm boys, a couple of Indians, and a real mixed bag of country boys. Most of them were younger and stronger than me because they'd come in right after high school. I was a college grad who had just recovered from pneumonia. Physically, I wasn't up to the demands they put on me. I went on a sixteen-mile, full pack (65 lbs.) jog, but near the end I weakened. The lieutenant grabbed me by the collar, choking me, and dragged me across the finish line. I then returned the favor and threw up all over him. I knew I'd end up back in the hospital if I kept this up.

The next day, I stood out there again with a full pack, about to go on another run, dreading each step. Someone handed the first sergeant a paper. He glanced at it, then called out, "I need volunteers. Everyone who has ever played badminton. Front and center."

Everyone who knew anything or ever played badminton stepped up. The first one said he was on the high school badminton team, and another said he'd played since he was a little kid. A third said he was on the college team.

I'd played badminton in college with Bruce. We played as doubles partners, and we were runner-up a couple of times, but we never could win. The kids we'd played against from

Indonesia were unbeatable—it was their sport, their game—but we learned to play well against them.

I thought: *I've got to make this good.* I stepped out, and said, "Florida State champion, Sir," figuring they wouldn't have time to check. They picked me and one other guy to represent the company.

I played my first match in a T-shirt, fatigues, and Army boots. I was playing for my life, I figured, playing to survive. No matter what, that bird was not going to hit the floor. Badminton saved me from more than the rigors of boot camp because I was on the verge of ending up back in the hospital.

I won that match, and kept winning. This is how it worked: I played; I won; I went to another camp and won for the battalion. I eventually won the First Army championship, which is for the whole East Coast. I didn't have to do any other marching, running, or push ups and could stay in bed as long as I wanted. In the Army, even today, an athlete is king. I even had a sergeant bringing me food in bed, and a major driving me to the matches. For four weeks, I did nothing but play badminton. As my health returned, I began to feel much stronger.

One day, a captain of the Third Army based in the Orient was scheduled to compete with me. If I beat him, I would do nothing but travel all over the world playing badminton for the next four months of my Army career.

The match was set to begin at five o'clock. Apparently my opponent, an officer, was due to arrive from Japan. Five o'clock passed, and he didn't show. By six o'clock the match was forfeited to me; I was officially declared the winner. I stayed to play with other people, just to practice. All of a sudden, at about seven o'clock, the officer arrived. He apologized and explained that his plane was late and he'd come all the way from Japan. He was a captain and looked old—probably all of 36, but he really looked old to me.

"Think about what you want to do, and let me practice a little," he suggested.

I watched him practice and he was hardly hitting the bird. *How did this guy even get here?* When he stopped, he said, "Will you forgive the forfeit and play me?"

The referee and the official were still there, so I agreed to play, thinking that this old man wouldn't last too long. As soon as I got out on the court I realized I'd been had.

I played three sets and won about four points the entire time. He took me three to nothing. I thought of Bruce and what he'd said at Stanley's Ping-Pong game: "Never give a sucker an even break." The officer had set me up by making it look like he wasn't very good and furthermore I had made wrong assumptions about his age. I had to learn this lesson again the hard way: *things aren't always what they seem.*

My badminton career came to an abrupt end and I was back to being a real soldier. I returned to the same company, that mixed bag of young guys from Maine. There were two black friends who really wanted to be in the Army, one whose eyes moved around in his head so that he couldn't focus. To be in the Army, you had to become a rifleman and pass the final by at least hitting the target. Unlike my skill with a bow and arrow, I was a sharpshooter who hit the bull's eye consistently. This fellow had to pass the final, but couldn't see well enough to hit the target. He was beside himself with worry about being dropped from the Army.

Ever the creative problem solver, I told him, "I'll go to the rifle range and just be there practicing. Pick a spot to be retested, two, three booths away from me. When you shoot, I'll shoot your target." So I shot out his target, and he stayed in the Army.

Another time, they had a war game supervised by a general, where they divided our battalion into two armies—a red army and a blue army. Each side was given a password. When the "war" started, you had to go through the woods. By this time, because of my recycling I had been in boot camp for almost 12 weeks, and I'd had enough. I went into the woods, found two nice bushes, hid between them and lay down to go to sleep. I put the helmet over my eyes and rested my head on my pack. All of a sudden, passing close to me, I heard the opposing army asking each other for the password and I heard what it was. I grinned. *This was just too easy.*

I got up, put on my pack, and made my way into their camp by giving the password. In order to win the war you had to put your helmet, like a grenade or bomb, on top of a pile of boxes. You had to get right up to them to do it.

They had two guards protecting these boxes. One was a young, very tall and athletically-sculptured black man. I walked up to him and gave him the password. "I'm not supposed to move and nobody is supposed to come within five feet of me," he announced. "Get away, or I'm going to hit you with my rifle butt."

"I'm supposed to take your place as guard," I said, but I couldn't con him. He wouldn't leave. I started arguing with him, all the while getting closer and closer to the boxes. I took off my helmet and put it under my arm. Still arguing, I put my helmet on top of the boxes. Bells went off. The games were over.

Feeling like a big shot, I walked around telling everyone I was the guy who did it. The general got up and said, "What kind of idiots are here? You let some *little shit* walk right up and put his helmet up!" It was pretty embarrassing after I had strutted around telling everybody I was the guy.

Near the end of boot camp, the young Indian who slept in the bunk below me started to act up. He was a reservation Indian, only about 17 years old and far removed from his cultural environment. As time wore on his behavior grew more and more erratic. One night, about a week before we were getting out of boot camp, I heard a scream. It startled me so much that I sat bolt upright in my bed. The next thing I knew, he thrust a bayonet through the four-inch mattress above him about where my shoulder blades had been. If I hadn't jumped up from the scream, he would have skewered me to the mattress.

I insisted they change my bunk assignment. About four days before the end of boot camp, the Indian went AWOL. About a month later, when I was reassigned to Fort Devon in Massachusetts, we heard that he was found wandering around Atlantic City on the boardwalk, completely out of it.

I pushed to get assigned to Fort Devon because a lot of my friends who were in my first company (before I was recycled) got themselves into the Signal Corps at Devon. My friends tipped me off ahead of time about what to say when asked what I did for a career, so I met the question with a ready answer. I said I was a photographer at weddings and Bar Mitzvahs.

As good as the Signal Corps was, the photography department was better because they never worked on weekends. I would be able to drive home on weekends from Massachusetts to see Judy, so this would be terrific.

I managed to convince them I was a photographer and they posted me to the photography department at Fort Devon, where I had friends to help me out. Even though they didn't normally work on weekends, I had to come in on my first Saturday there because a general was coming to the base. They were having a parade for him, and wanted pictures for the camp newspaper.

They gave me the camera and I couldn't very well tell them I didn't know what I was doing, so I just did it. It was a big camera, the kind that uses film plates. I figured out how it worked and took the shots. When I brought the camera back with my fingers crossed that the shots came out, the sergeant in charge, who'd sent me on the assignment, told me to develop the pictures.

Luckily, I had developed rolls of film before. In the basement of our Beechhurst house I'd even had my own darkroom, but this was different. These were big plates of film, not rolls. I didn't know what I was doing, and I lost all the pictures. It was the first time a general had ever been to Fort Devon, and they had no pictures for the newspaper. The sergeant was furious. My friends in the photography department calmed him down, and I was reassigned to a job where they felt I couldn't make a mistake. I took pictures for I.D. cards. The subject stood on a designated spot and smiled, and I clicked the shutter. That's what I did all day long.

With about four weeks left to go on my six months' tour of duty, I had KP duty one weekend. Ever since being assigned to Fort Devon I had been driving home every weekend to see

Judy with some of the other guys from New York. Now I was stuck.

What everyone did to get out of KP duty was pay the company clerk ten bucks. He split the money with the cook who marked you present, but you didn't have to do KP, you could just slip away. My friends had done this before me, and we heard it had been going on for years.

I was nervous. I knew you can never go wrong by doing the right thing. Whenever I did something wrong, I always got very nervous and my hands would sweat. I had played hooky only once in my life as a kid and now felt just as nervous then. But my friends urged me, "Come on. We did it and it was fine. It's ridiculous not to." I paid my $10, and drove home for the weekend.

I was home early Saturday morning, still asleep, when I got a call from one of my buddies telling me that I had been reported AWOL. I knew he wasn't fooling; nobody would kid about this. AWOL was a serious charge, and I could end up in the brig (jail). Shaken to the core, I knew from that moment on that I would never again do something that I knew was wrong just because everyone else did it and got away with it. I had a big problem: I had driven all the guys and had to take them back.

I waited until the end of the weekend and drove the others back to camp. If I got caught even going into camp and was reported, I could be arrested. I might have gotten sent away for up to a year. *How could this happen so close to the end of my tour? I was supposed to be getting married ten days after I got out of the service!*

On the way in, I hid in my trunk, covered with blankets. The guys drove my car up to the guard gate, had their I.D.

checked, then drove through. If the guards had checked my I.D. I would have been arrested on the spot.

I slept in my car all night. When dawn broke, I went to the first sergeant's office and reported that I was AWOL. He was a career first sergeant who had obviously been around, so I found it hard to believe he didn't know about the deal with the cook. My friends had suggested that I should tell him about the payoff, but having grown up a streetwise kid in the Bronx, I didn't think that was the right way to handle this.

"Why were you AWOL?" the first sergeant asked.

"Sir, I'm getting married soon. I went home to see my fiancé." I didn't tell him about the payoff, or the company clerk, or the cook. Nobody had turned me in. It was just my luck that, after it had been going on for three years, the whole scheme blew up on the very weekend I chose to go along with it. The company clerk had gotten married and the cook was sick and didn't come in.

"Did you pay anyone off?" he asked, a dead giveaway that he knew.

"No, Sir."

"Okay, go in and talk to the captain." The first sergeant looked at me squarely. His question and his look verified my suspicion: he was probably in on the scheme and was only sending me to the captain because he knew I wouldn't turn anyone in.

I spoke to the captain, giving him much the same story. I was frightened about what would happen to me, but I knew I had done the right thing. Two hours later, I received my punishment. I was placed on Article 15—two weeks of regular daytime duty, with restriction to barracks every night and all weekend. It could have been a lot worse, so I was relieved.

But as luck would have it, I didn't sit around with nothing to do. There was a baseball field behind our barracks where they held a practice game between a made-up team and the camp's touring baseball team. The other team was short players, and I was sitting around the barracks, serving my restriction. Agreeing to play second base if someone gave me a glove, I went out to the field in my boots. It was the best baseball I ever played. I got five hits, and made some amazing catches all over the field. They wanted me to join the camp baseball team. I just wanted to get out of the Army as fast as I could.

As I reflect back, I feel the Army prepared me for life ahead and put things in perspective. It gave me an appreciation of the simple things in life, such as: going to the toilet when you wanted and not sitting shoulder to shoulder with thirty other guys; showering at your choice of temperature; sleeping in a normal bed; coming in out of the rain or using an umbrella; going to sleep and waking up when your own body dictated; and eating chicken without potatoes and apple pie on top of it.

CHAPTER 9

Judy and I
Begin Life Adventures

I got out of the army on September 17, 1957. I was about to begin life's adventures. *"Life is usually simple multiple choices: spectator or player, love or hate, create or criticize, give or take, right or wrong."* — *Donaldism.*

Judy and I were to be married on September 28th. Meanwhile, my father had sold Rexart, so I no longer had a job. In ten days I would have a wife, but no job.

We had a lovely, formal wedding with a Rabbi. The reception was held at the Savoy Plaza in Manhattan (since torn down), in the area of 58th and Fifth, where the General Motors building is today, across from The Pierre and The Plaza. There were about 120-150 guests. Judy and I had two tables of friends there. Most of the people were my parents' friends and Judy's relatives from Florida.

It was a big traditional affair; we posed for the wedding photos, cut the cake, and did all the usual things. Roger was the best man, and Bruce, Stanley, Howie, and my cousin, Gary (Ann's brother) were ushers. Judy's older sister by seven years, Bea, was her matron of honor, and her other sister, Sandy, was her maid of honor.

That night, our wedding night, we stayed in a suite at the hotel. The next morning we were leaving for our honeymoon, but first we were supposed to meet my parents for breakfast at Reuben's.

In those days, your first night was really your *first night*. We went up to our room in great anticipation. I got out of my tuxedo, into my pajamas, and climbed into bed to wait for Judy who was in the bathroom. I waited and waited. The next thing I knew, I wasn't waiting . . . I was sleeping. When Judy came out of the bathroom, I was sound asleep, and she didn't wake me.

About three in the morning, I woke up to the sounds of sobbing. Remembering what I was there for, I reached out for my bride and we consummated our marriage. The next morning, when we met my parents at Reuben's, they couldn't get over Judy's radiant glow.

As a wedding gift, my parents gave us our honeymoon trip—island hopping in the Caribbean by plane; we planned three-day stops on each island over a two-week period. The weather was great, but the political climate was not as sunny.

We flew first to St. Thomas in the Virgin Islands. The only trouble was they were having problems, and you weren't allowed to leave the hotel at night and go into town because there was a lot of crime. My father-in-law had given me a German movie camera, and I became a director. Starting with St. Thomas I must have taken a dozen reels of film, but like the Army, not one picture came out.

From St. Thomas, we went on to Haiti. There was fighting in Port-au-Prince because Papa Doc, as Duvalier was called, had just taken over the country. We stayed in a luxury hotel, but instead of sleeping in the bed, we spent the first two nights on the floor, listening to the gunfire outside. It was really kind of exciting. Unafraid, we went into town in the daytime. We were so young we didn't even realize the danger; we felt impervious, detached from it.

Most of my life I felt impervious to danger. Perhaps this feeling was the character flaw that helped to set the tone for some of the things that happened in the future.

In Haiti, we bought some of our first art: the beautiful, bright watercolors that Haiti is known for. I looked up Haitian friends, two brothers who were in college with me. They were ping-pong players and tennis players, and I had gotten to know them very well. Their family owned a large retail store in Haiti, and now they were running it. They gave us great discounts on some art and a wooden sculpture by Haiti's most famous artist.

One day, they took us around Port-au-Prince. That night, we had dinner with them and an American shopkeeper they knew. They told us how bad this Duvalier was and explained the terrible situation. They asked me to do them a favor and carry some papers out because they couldn't send any mail out of Haiti. Haitians couldn't bring anything out of the country because Duvalier was going through everything and restricting it. The American inferred that he worked for the U.S. Government, but whether he did or not, I don't know for sure. He talked like he was CIA.

They handed me a huge folder of papers, which I put in my bag and took through customs. I felt only like I was doing my friends a favor, not like I was doing anything dangerous. As a U.S. citizen, it didn't occur to me that anything could ever happen to me. We left Haiti without incident, and flew to Santo Domingo.

We chose Santo Domingo because the World's Fair was on at the time. We arrived in the afternoon and the next morning went to the fair site. Judy and I were the only two tourists walking around this huge fair. The rest of the people were soldiers with machine guns over their shoulders. Trujillo had mounted a revolution, and nobody was out in the street. Without people, this World's Fair was like a ghost town. We didn't stay long.

From the Dominican Republic, we flew to Cuba, where we met a lot of people. There was a revolutionary in the mountains attacking people; it was Castro. Although the Cubans weren't thrilled with Batista—I used to hear stories from shopkeepers who told us how bad Batista was—they were far from happy with soldiers carrying machine guns all over Havana.

One night, we went out dancing at the Tropicana nightclub, which was really an event. It was a "Busby Berkley" outdoor extravaganza staged in a tropical paradise with hundreds of beautiful, elaborately-costumed women, many dancing in a synchronized fashion—a breathtaking sight among the palm trees and colored lights.

We ended our honeymoon in Miami, the first place we'd been in almost two weeks without some kind of revolution going on. I mailed the papers the Haitian brothers had given me, without looking at them and without thinking much of it.

On our second day in Miami, we picked up a newspaper, shocked by the headline. Two brothers, former University of Miami students, had died in Haiti—one had fallen down the stairs of a police station and broken his neck, and the other had a heart attack. They were the same two brothers we had visited. At that point, Judy and I started to realize that we had probably been on a pretty dangerous mission.

We returned to New York without mishap. We had arranged to rent an apartment in LeHavre in Whitestone in a brand new apartment house being built by Levitt. When we got back from the honeymoon our apartment wasn't ready to move into right away, so we rented the cheapest place we could find, a little furnished apartment in Long Beach. I was looking for work, but Judy was already working in the city. Every morning, we walked the seven blocks to the train with the wind whistling off the ocean; it was freezing. Meanwhile, my mother and Judy shopped for furniture, preparing for the day our new apartment would be ready.

I remember the first meal my new bride cooked for me. As Judy carried the broiled chicken to the table, her hand was shaking so much that the plate rattled and the chicken hit the floor. I picked up the chicken, wiped it off, bit the bullet and ate it. Naturally, Judy was upset, embarrassed and disappointed, but I calmed her down and told her how good it tasted. Although she was nervous about her cooking at first, Judy became a confident and proficient gourmet cook, as good as or better than my mother. Having eaten many meals cooked by my mother-in-law, I never imagined this would happen.

Pretty much a natural, Judy learned to cook by feel and taste, as well as by reading cookbooks and recipes. She built a library of cookbooks and still loves to read them. My mother gave her some family recipes and an old-fashioned cookbook, which she still has.

Judy cooked basically whatever I wanted to eat, but she put her foot down after she made rabbit once. As a joke I started to twitch my nose at the dinner table, and never again would she cook rabbit for me.

As newlyweds we went out to eat, but couldn't afford anything fancy because we were just making ends meet then. We went to movies together, and later, when the kids were little we took them in the car to drive-in movies with us.

Our new apartment in LeHavre was in a youthful community that had a swimming pool, tennis courts, and a waterfront restaurant—everything a young couple could want. Iris and Stanley Rabinowitz had rented an apartment in the same complex and so had Judy's sister, Sandy, who had just married Norman Berlin. We already had friends in the complex and everyone there was about our age.

The apartment suited our newly married needs perfectly. It had a large living room with a terrace and big spans of glass. It was in a very nice area near the Long Island Sound, where I had sailed as a boy. Mitchell, our first son, was born there, and it was a great community for bringing up children. But the apartment only had one bedroom and the closets were like freestanding boxes. After three years, we outgrew the place and moved to Jericho.

A couple of interesting things happened while we lived in the apartment complex in LeHavre. Of all our friends there, Iris was the first to get pregnant. I was assigned the task of being the designated driver. She was going to Doctors' Hospital in New York City, so it was a drive. When the big day arrived, Stanley got in the passenger seat and I drove while Iris was in the back seat, moaning and groaning. It made me so nervous that I couldn't drive. I pulled over. Stanley finally had to drive me, along with Iris, to the hospital.

Somehow I managed to drive Judy to the hospital when she had each of her three babies. One day, when she was pregnant and not supposed to eat spicy food (which she loves and I can't eat at all), I happened to taste her pasta. Wow . . . it was so hot it almost put me through the ceiling. I took it away and she started to cry. For the rest of the pregnancy, and the next ones, I became the food taster.

Our first son, Mitchell was born in Booth Memorial Hospital in Flushing on December 15, 1959. He stayed there for eight days, and had his bris. My mother and dad came to the hospital for the bris, but there was a bad snow storm that day, and their car skidded coming out of their driveway. They smashed up a new Mark IX Jaguar that they'd bought in London and had used to tour all through Europe. Eventually, after it was repaired and years later, they gave Judy that car. It was unique because it had tables in the back. She used to take the kids to McDonald's in this big fancy Jaguar and they would eat hamburgers and fries on the tables in the back of the car.

Getting back to our first year of marriage, we began our art collection by buying our first real piece of art at a charity affair. Aunt Ruth pushed Judy and me into buying a $350 Picasso woodcut—a very small one. This sum was five weeks' salary for the job I had then. It was a *huge* amount to spend, but after all, it was Picasso. We still have it and treasure it.

In the second year of our marriage, our last year in the apartment, I got a call at work from Judy that she wanted to go to the hospital because she needed a surgeon. When I asked her why, she explained that she got her finger stuck in a silver salt shaker while polishing it and couldn't get it out. Instead of going to a hospital, I suggested she go to a jeweler. They cut the salt shaker off her finger.

When Judy and I first moved into our new apartment, I was still looking for a job. I mentioned that my mother had always thought I should *not* be a businessman because I had no business acumen. At the beginning of my career, I began to wonder if she might be right because the only job I could get was selling encyclopedias for Britannica, and the only set I sold was to Iris's mother, May Kessler. The minute I came to the door she wanted to buy it from me, but I made her wait so I could go through my whole spiel. I lasted less than a month with Britannica, and then left.

During Christmas, I worked for Altman's, then Judy was able to get me a clerical job at Gillman Paper Company in Rockefeller Center where I worked for six months. It was boring as the day is long. At the time, Judy had a secretarial job paying $90 a week. Together we earned $170 a week and fortunately our rent was very low, about $135 a month. I don't believe my parents were helping us out financially.

My father eventually helped to get me a position with Pierce, Mayer, and Grier, a mortgage and brokerage firm that also did real estate sales. He had used them for his building financing. My new office was located on 42nd Street in Manhattan, and I was working for the best. Henry Pierce was the most incredible salesman, a legend in his time who could sell you the Brooklyn Bridge . . . twice. I met with him every week.

Working at Pierce, Mayer, and Grier, I made a lot of friendships that I would carry through my years in the real estate industry. One of them was Donald Zucker. At that time I was a loan solicitor and he was a placer, which meant he went to the banks and got the loans while I went to the buildings and got the business.

In my first two and a half months I made $6,000 in commissions. That was the first business success I had met with and I was very happy. The only problem was that we were going into the Eisenhower years when money dried up like powdered sand and there was nothing available for the next six months. It made no sense to even call a builder because there was nothing you could do for them.

I had a salary against a draw, so after six months I had begun to seriously eat into my $6,000 with no more coming in. One day, I found myself sitting in the movies in the middle of the afternoon, and looking around, I saw a whole bunch of losers. I stood up and walked out, knowing I had to get myself out of this. I couldn't sit around and do nothing and I didn't have enough confidence to stick it out.

I heard Cohen, Hall & Marks, the largest manufacturer of men's piece goods at that time, had sales jobs paying $95 a week. It was a good corporation, so I applied. They told me in order to get a sales job I would have to go through their testing. For a week I took I.Q. tests, inkblot tests, psychological tests, recognition tests, multiple-choice tests, and had two interviews. They called me in the following Monday to tell me: "Something wild has happened in the test scores and we want to retest you to make sure we haven't made a mistake. You have the highest score that anyone has ever gotten, including all our executives, our marketing staff, and our chairman."

Not only did I have to take all these tests again, but I had to have interviews with their sales manager, their vice-president of marketing, and the chairman himself. When I told my father, he said, "Are they hiring you for a $95 a week job or are they hiring you for president?"

Finally, after all that, they called me in the following Monday. "We are not giving you a sales job," they announced. When I asked why, they said, "Because you scored the highest in marketing—that's retailing—and we have a retail chain. We want to offer you an assistant manager position at a Robert Hall store."

Robert Hall was the first chain of discount stores for men's suits, what in industry terms was called a "bare pipe rack." They explained that I would start off at $110 a week, and after six weeks in their training school, I would become an assistant manager at $135. In four months I would be managing my own store and making $155. After that I could become a district manager at a substantial raise . . . and who knew what after that.

I had no plans to be a retailer. Granted, I knew retailing because my grandfather and my father had been retailers. It seemed like a good offer, a way I could make a living and have a future. I knew it could be a good living, but I also knew only too well that a retailer worked *long* hours. I told my father about their offer, and that I would have to go away for six weeks training.

"Listen, son, if you want to be a retailer, don't work for anyone else," Dad said. "If this is really what you want to do, I'll set you up in retail."

"Well, it seems like the thing to do," I said.

We looked around for upscale locations to retail gifts—china and flatware—and decided Manhasset was our best choice. Friends of my father owned a store called Porch and Patio, to whom he'd sold Rexart's aluminum furniture. They had too much square footage, so we arranged to share the space and help them pay the rent. On one side of the store, we opened up a separate department with our own register.

By now, Roger had completed his freshman year at the University of Miami. In his second year, he dropped out of college and came into the newly opened Gift Imports business as my partner. We also hired a pretty young lady named Lee who ended up marrying Roger.

The idea of Gift Imports was born in 1959 and became very successful. We sold what my father had always sold: gift ware, china and flatware. Although I had followed this direction because of a test, rather than having made a conscious career choice, I enjoyed the years I worked as a retailer.

We were in the Americana Shopping Center, and our goal was to "sell in the neighborhood." For people living on Long Island this neighborhood was their equivalent of Madison Avenue. We became the most upscale shop in the center, and that's exactly the direction our landlord, Frank Castagna, wanted to take that center. Now a lifelong friend, Frank visited my store every day, and loved what we were doing.

My grandfather, who had started out with a pushcart on the streets of New York, always said: "You can never do business from an empty cart." My father held the same philosophy, and imparted it to me. Because he had been in the business he was able to establish credit for us. We stocked as many sets of dinnerware as any department store, and had a large and varied selection of flatware, silver, and gifts.

It turned out I had talent: a great eye and a good feel for what people wanted on an upscale level. I became a very good buyer, going to the city three to four times a year to fill up the whole store with merchandise. Eventually, we did more retail sales per square foot than was reported by the largest companies in non-urban areas in the country.

We had been open about six months when I learned a valuable lesson. We'd hired a manager that my father and I both liked. Ernie didn't have any references, but he was very personable and hardworking, probably too good to take the kind of salary that we could afford. He was there over a year before we realized something was wrong, something didn't add up with our layaways. We baited a trap for him. The store offered a layaway plan where people could put a deposit on something and pay it off over time. We discovered that the initial deposit would get into the register but the payments went into Ernie's pocket. If something looks too good to be true, it probably is.

Our next manager, Bernie Gafcovich, had escaped from Castro's Cuba. He began with us as a stock person, and eventually became a manager. Bernie was as hard working a person as you could ever hope to hire. While he was working for us, he took five days off to go back to Cuba and blow up his father's clothing factory. It had been taken from the family by Castro, as was their house and everything else they owned. Bernie eventually moved to Miami and became one of that city's largest retailers; he had a very successful chain of children's stores.

Our mantra at Gift Imports was that we would sell the best merchandise at the lowest prices. Today, everybody says that, but in 1959, prior to discount stores it was a new concept. We kept my grandfather and father's old code for coding prices, made up of Grandma Rechler's maiden name, Goldstein. "G" was a one and "O" was a two and "L" was a three . . . an "X" was ten. That was the coding on every price in the store.

Out of thousands of items, there wasn't a retail price I didn't remember or the amount I'd paid. When it came to numbers, I had almost a photographic memory, like a computer. In later years, I felt I had used up my memory bank by filling it with all those numbers and prices.

We were at the Porch and Patio location about three years, and it was such a success we decided—urged along by my landlord, Frank Castagna, and my father—to move to a larger store. A 6,000 square foot store, one that is really large, became available in the shopping center. It went through the center front and back, and had two floors. Since we were doing so well, we decided to take it and expand the products we sold.

My father knew a jeweler, Irwin Pearl, who was a major designer. We agreed to be partners with Irwin in a jewelry department. I would do some buying with him, but he would man the area and decide what stayed and what he would trade out. He would also sell some of his own merchandise that we would buy from him.

We added a luggage department and expanded our other departments. We did more advertising, more mail order, and now we were drawing from all over Long Island. Even people from Westchester came to our place. The new store was a huge success.

We opened another Gift Imports store on Sunrise Highway, but it was a small store that carried mostly silver and gifts. It was on the south shore of Long Island, about a 45-50 minute ride from us. We didn't do well there. What I learned early on was that it is hard to run something when you're not there (absentee ownership). I would later base business decisions on that lesson for most of my life.

Rexart had been sold, but Dad and Uncle Mort and Uncle Jack had stayed on to run the operation on a three-year contract with the new owners. The first conglomerate, U.S. Hoffman, that now owned Rexart, was run by a math professor named Marcus.

One day, Marcus, a supposed genius, called my father and uncle into his office for a meeting. He told them profit wasn't important. Volume was the key. They were to increase the volume, and not worry if they lost money doing so. Always a good businessman, my father understood that this approach would not work. All it would do was hike the company stock. They had taken mostly stock in payment for their business, and they realized they could eventually be in deep trouble if they followed this advice.

The stock they had taken at $20 now shot up to $35, but they were locked in and not allowed to sell for another year. The difference at that time was that there was no actual law against selling it. (Shortly after this U.S. Hoffman debacle, the federal government passed a law.) However, they could be sued by U.S. Hoffman for contract violation for selling the stock so soon.

My father had always been lucky. He transferred the stock to a banker friend who was able to sell it all over the country for $32 instead of $35. As they were selling the stock, U.S. Hoffman was buying it back to protect their price. They didn't know where it was coming from. By the time they finally discovered what was happening, the stock had started to plummet and they had to get out. They had built themselves a number of factories for their own businesses, and at that point decided they would go into the business of building industrial factories.

My father and my two uncles, Morty and Jack, all did amazingly well. They actually bought swamp land, New Town Creek, at the corner of Queens and Brooklyn in Maspeth (after doing their homework), and turned it into a highly successful industrial park, the first in New York City. My father always had that kind of luck.

After my father finished developing the Maspeth property (the industrial park was called New Town Creek), they continued with this kind of development, growing more and more successful. Uncle Mort was the builder and my father and Uncle Jack were great salesmen. Everyone liked them. They played off each other well. Dad had a similar relationship with his next partner, Walter Gross, a long time friend, when they built Vanderbilt Industrial Park. He and Walter would play bad guy-good guy. Using this strategy my father was always the good guy, and either Uncle Mort or Walter the heavy. None of them had to stretch too far out of character to play their parts. They were tough but honorable.

Many years earlier, my parents had met Walter and Barbara Gross, who became their traveling companions and best friends. When my mother pushed my father to break away from my uncle, he became partners with Walter Gross, who was in construction. They bought 400 acres from McKay Radio, the property used as the "Voice of America" during World War II, and built Vanderbilt Industrial Park (VIP) in Hauppauge, Long Island. It became the second largest industrial park in the United States, and that property is still in the family's portfolio today.

After they finished New Town Creek Industrial Park, my father couldn't bring himself to negotiate with his younger brother, so he asked me to do it. I suggested a way of splitting up their holdings. Like the National Football League draft, Uncle Mort would pick a property, then my father would pick one, then Uncle Jack.

The system worked very well. The best properties were obvious, but as they got further along, my father, who had done his homework, took the ones that didn't look good on paper. One such building was the Fuller Brush Building in Farmingdale. It was on a 30-year lease with the rent going down every five years, and it didn't look like a profitable building. Having visited each building and being friendly with the principals, Dad had talked to everyone and found out that Fuller Brush planned to move out and give up their lease. The lease would be worth much more when it was re-rented.

I negotiated for my father. After taking a couple of other buildings, we purposely were left to take that building and it was a home run for him. I felt good about being able to help in that process. It was as if a huge weight was lifted off him and he became himself again.

We had been in business for three years with Gift Imports when my mother died suddenly in the summer of 1962. I was in the store when I got a call that she had collapsed while shopping in a supermarket and had been brought to Roosevelt Hospital. They told me on the phone they felt she had overdosed. "My mother doesn't take drugs," I said. Roosevelt Hospital was a notorious drug hospital, so that's probably all they could diagnose.

I drove over there immediately, with my father in the car. Picking my mother up, we put her in the car and headed home, planning to call our doctor to come to see her there on an emergency basis. She was writhing in pain in the back seat, and while we were driving home, she said, "Donny, I'm not going to live through this."

My mother had never been sick a day in her life. "What are you talking about? Of course you're going to make it," I told her.

"No, I'm not," she said.

We got her back to the house and called our family doctor, Dr. Henry, who lived nearby in Great Neck. He was over in 10 minutes. When he got to the doorway of her bedroom, he said, "Call an ambulance. Give me the phone."

In about forty minutes she was in an ambulance and they rushed her to Long Island Jewish Hospital. She had an aneurysm in her brain. We got the best neurologist that we could get, Dr. Bender, and the best neurosurgeon, Dr. Epstein. We had round-the-clock nurses. I was there almost all the time. Roger and I would go at alternate times so we could cover the store. My father had a difficult time dealing with her pain and all the tubes. My mother slipped in and out of consciousness for eleven days.

The night shift was a problem because the nurses would leave her alone. I sat by my mother's side through the night, even though visitors weren't technically allowed to stay overnight then. One of the nurses had me physically thrown out of the hospital. Knowing she had disappeared the night before, I waited about an hour, then sneaked back in. Mom had so many tubes, and I knew if she lost any of the tubes in her it could be fatal. About three hours later, the nurse returned and I told her she was fired. She got some guards to take me out.

"You're going to have to fight me to get me out of here because this woman left for three hours. And she did the same thing last night," I told them. It was only then that they got special permission from the hospital to let me stay overnight. I learned two things from that: don't take anything for granted, and a hospital is no place for sick people. If you have to go to a hospital, you should always have someone with you to run the show.

After eight days, my cousin, Ann, and my Aunt Ruth came to the hospital. I waited outside while they visited my mother, and after they left, I walked into her room. Mom opened her eyes and said, "Ruth means well but makes me so nervous." She closed her eyes, and those were the last words she ever said. She passed away that day.

My father went through a very tough time then, and for a while, his grief was debilitating. He pulled back from his real estate partnership with Walter Gross. One of his great loves was art, so he opened up a hobby gallery on the lower level of Gift Imports. He carried Impressionists and the Ashcan School.

Two things came out of my mother's death that I had never heard before. One, was that when my parents had taken an around-the-world-trip with Walter and Barbara two years previously they had met a fakir in India. Fakirs were palm readers who read your future and your fortune. He read everybody's palm, then he called my father and Walter aside. He told them that my mother had a very short time to live. They laughed it off, but never told my mother. I heard the story after she died.

The other thing I heard was from my grandmother who told me about my mother having polio as a child and about the doctor who'd told her that Gloria would not have a long life. My mother died at age forty-nine. The two predictions had come true.

I was twenty-eight years old and Roger was twenty one when our mother died. She had become very much a part of our son, Mitchell's, life; he used to play piano with her. Mitchell was three and a half when we told him his grandmother wouldn't be coming back. While we were sitting Shiva at my parent's house in Kings Point and my father was so upset and crying, Mitchell went around the house and tore up the pictures of her. It was a release of his young anger until we explained what had happened and helped him understand.

A month after my mother died, our second child, Glenn, was born August 6, 1962 and named for Gloria. Again, I drove Judy to the hospital. By then, we had moved to Jericho, so Glenn was born in North Shore Hospital where Judy only stayed for three or four days.

Each of our three sons was named after their grandmothers who had just died. Judy had lost her mother early too, just two months before Mitchell was born, and he was named after her. My grandmother, Mildred, died just before our third son, Mark, was born and he was named for her.

CHAPTER 10

Young Family and Career

In 1960, Judy and I bought our first house at 263 Birchwood Park Drive in Jericho. It was a split-level, three-bedroom house with a finished basement. The community was about ten years old at the time we bought the house for $27,000. Judy and I decorated it ourselves, the first time in my life that I'd done my own decorating. I had access to a wide range of items, since I was buying for Gift Imports, our store, and imported a lot of products for interior designers. Unlike the apartment, which had been decorated by Judy and my mother, the house was a good mix of both our personalities.

In the first year we lived on Birchwood we got invited to a party with our friends, Debbie and Arnie Widder, on Friendly Lane, just around the corner from us, unaware that these parties had a notorious reputation as reported in *Confidential*, a national scandal magazine. Typical of suburbia in the early sixties, there were swingers' parties (partner-swapping) going on at Friendly Lane. The street name made it a doubly good story for the tabloids, and according to the report, due to divorce many people had already moved out of the neighborhood—some just exchanged wives and homes. We really didn't know the people who invited us to this wild party, but we were curious to see what was going on around the corner.

The place reeked with the smell of pot. It was a very fast crowd and there was some touchy-feely going on and jokes about throwing our keys in a pile on the table. After an hour, I told Judy, "Come on, let's go. It's time to get out of here," and the four of us left. We never went to any parties in the neighborhood again.

Friendly Lane, like Birchwood Drive, was home to many different kinds of people. That first year, Judy had met one of her best friends, Debbie Widder, a woman her age from Boston who lived on Friendly Lane. Debbie had two children the same age as our boys.

Our first experience with the anxiety of parenting started early. The first time Mitchell tried to crawl out of his crib as a baby he fell on his head. When he was eighteen months old (I was twenty-five at the time), he came down the stairs of our house in Jericho at 10:00 P.M. As it was a split-level home, the stairs, which were covered in asphalt tile with a metal edge, consisted of three short levels. Mitchell slipped and ripped his whole lip open. I picked him up, and Judy called the doctor who told us to go to the emergency room at Plainview Hospital.

It was 10:30 at night and we were waiting for a doctor in the emergency room. We saw what looked to us to be an Indian exchange student walking down the corridor, dressed in surgical garb, flipping a coin up in the air and whistling. As he came closer, I wondered how qualified a surgeon he could be. I was afraid Mitchell's injury could turn into a cleft lip. The "doctor" advised us that the sooner our child had surgery the better his chances, but I wasn't going to give Mitchell up to this man. I made a quick decision. Judy and I took Mitchell from the hospital and drove to the doctor we knew who had an office in his home.

Mitchell never stopped crying the entire time. Our doctor looked at him, and decided his injury was bad enough to bring in the best child plastic surgeon on the island. (This surgeon later became the head of Pediatric Surgery for Children's Medical Center on Long Island.) When the plastic surgeon finally took him it was about 1:30 in the morning. He was a huge man—maybe 6'4" and 300 pounds—with huge hands. As soon as the doctor held him, Mitchell stopped crying. It was amazing. For the first time all night, I felt confident. The doctor did a terrific job repairing the damage, and by two o'clock we were headed home with a sleeping child.

I was home alone with Mitchell in 1962 when Glenn was born and Judy was in the hospital. About three o'clock one morning, Mitchell woke me from a deep sleep. He was holding a hammer and told me that when the baby came home he was going to pound it in the head. "You can't do that," I told him firmly. "He is your little brother. You've got to take care of him." I put Mitchell back to bed, but I couldn't sleep for the rest of the night. It had shaken me to wake up and find a three-year-old standing over me with a hammer, and then threaten to hurt his baby brother.

A few days later, I picked up Judy and the baby at the hospital, and brought them home. We put baby Glenn down on the daybed in his room where the nurse was going to sleep. Mitchell crept into the nursery with his hand behind his back.

I stood over Glenn, shaking and ready to pounce. Mitchell looked at his new baby brother, took his hand from behind his back and gave him his favorite stuffed animal.

There was almost a three year age difference between the boys with Mitch being born in December of 1959 and Glenn in August of 1962, but they got along well all of their lives. (Actually, everyone gets along with Mitch because he is easy-going.)

For the first five years of our married life I was still in the Reserves and had to report for duty on my two weeks of vacation every year. Judy and I would go away for a long weekend only twice a year and that would be the total extent of our vacation. When I was in the retail business and neither of us had mothers to leave our young children with because they had both died young, we couldn't afford to go on a regular vacation, so we'd go on gambling junkets for the weekends.

Once, we went to Victoria's Sporting Club in London on a three-day junket, and we had trouble coming back. For some reason, when we went through customs they ripped our bags apart. On these junkets they expected you to gamble because so much of it was provided free. While we did do *some* gambling, I was careful because I couldn't afford to lose much. One day, on the London junket, we disappeared and took a day trip to Paris. It was the first time we ever saw Paris and we were in awe of it.

We took a gambling junket trip to Lucaya Beach Resort in Freeport, the Bahamas, with Stanley and Iris. One morning, I went down to the casino early when it was almost empty. I asked two bleary-eyed gamblers how to play craps, and after they showed me, I started to roll the dice. Others started coming to the table, but I held the dice for almost fifty minutes. The only trouble was I didn't really know how to bet properly.

With a streak of beginner's luck I won about $2,800. In no time, the place was jammed; people were standing on chairs around my table *betting on me*. One guy won $62,000, but I was thrilled with my big win. That day I established my reputation as a gambler, and with that reputation came more trips.

We flew back to the states from the Bahamas on the Flying Tiger Airline, a charter arranged by the resort. It was drizzling when we arrived at the airport, and I noticed before boarding that our pilot reached out the window to wipe the plane's windshield with a paper towel. It gave me pause for concern, but we boarded the plane anyway.

As we flew over Washington, we ran into a bad electrical storm and the plane began bouncing around like crazy. People were screaming. A metal food cart broke loose, hitting one of the stewardesses and knocking her down. She was lying in the aisle with a broken leg. Suddenly, the plane began to plummet. Luggage flew out of the overhead bins. I had my safety belt on very loose and I felt like a balloon on a string hanging above the seat as the plane did a nose dive.

An off-duty pilot happened to be sitting next to me, giving me a play-by-play description of what the pilot was going through. "This is bad," my seat mate said gravely. "This pilot has to be damned good to pull us out of this. He's working his flaps and trying."

I glanced across the aisle at Stanley, who was clutching the arm rests so hard his knuckles were white. Judy and Iris were crying. I just watched the whole thing happen, like an observer. It was a peculiar thing I thought later: my lack of nervousness.

The pilot did pull the plane up, of course, or I wouldn't be writing this book, and they took the stewardess away in an ambulance when we landed. Most of the passengers were still hysterical and screaming, unable to get over it even after they were safely on the ground. The gravity of our situation didn't hit me for a couple of days. I probably felt it wasn't our time yet.

After that, we went on one final junket with Stanley and Iris, a short vacation at the Dorado Beach Hotel in Puerto Rico. One night while we were gambling, I ran out of money and asked Stanley if I could borrow some. He loaned me $200, and almost immediately I won. Wanting to repay my debt, I handed him back his $200 in the middle of the crowded casino. The next thing I knew, two big guys ordered Stanley to come with them. Thinking we were washing money, they took him into the kitchen and threatened him with a knife. In those days, the junkets attracted a tough crowd; many of the casinos were staffed by Cubans thrown out of their country when Castro came in, Mafia really. That was the last junket we ever went on. We decided it would be cheaper and less stressful to pay for our vacations.

When we still lived on Birchwood, one morning about 6:00 AM we heard a knock on the door. I went downstairs, unable to imagine who would be calling at this early hour, and opened the door to see two policemen. "Get dressed," they ordered. "We're taking you to the station. Your registration is late." One showed me the handcuffs.

"What on earth?!" I couldn't believe it . . . handcuffs? Admittedly, I had a late registration—two days late—but I had reapplied and was waiting for a temporary.

I had no choice but to go with the police. They took me to the District Attorney's office in Nassau County in Minneola where they fingerprinted me, questioned me, and then put me in a cell with criminals. I still couldn't believe it. It felt like the old Abbott & Costello joke: as Costello headed for the electric chair, he asked Abbott to pay his $5.00 traffic ticket. I would have laughed, but didn't want to provoke my cellmates.

After awhile, the cops cuffed me and put me in a paddy wagon with a bunch of criminals going to the courthouse. In the paddy wagon, one guy said, "What are you here for?" A tough-looking character said, "I stabbed someone." Another said it was for a hold-up. One with a jagged scar on his forehead beat someone up in a bar fight.

Here I was in a paddy wagon, wearing cuffs just like the rest of the thugs. They asked me what I'd done, but I was afraid to tell them that I was two days late on my registration. I made up a story about getting into a fight.

We arrived at the courthouse and they took me inside. Still handcuffed, I sat and waited on a hard bench while the judge went through the cases. A couple of hours passed and it was now late afternoon. I began to worry that I really was going to jail. I hadn't been allowed to call a lawyer because I hadn't been officially charged yet. These proceedings were held so the judge could set the trial date and charge you.

They finally brought me before the judge. I told him that I was here because I was two days late with my vehicle registration. He glanced down at my cuffs and said sharply, "What is he doing here? Take the cuffs off him."

The judge stood up and screamed at the detectives: "Get the D.A. on the phone right now and tell him to come over here and bring every record on this man. And I told you to take his cuffs off. *Now!*"

The detectives scrambled for the handcuff key and the phone. The D.A. rushed over and the judge screamed at him. The judge ripped up the files, but charged me a one dollar fine so as not to give them any excuse to reopen the case. It was a perfect example of bureaucracy going awry. People get so caught up in bureaucratic red tape that they don't know how to think for themselves anymore.

It was almost a year after my mother died that we opened the large Gift Imports store. My father was wrestling with his grief by going to bars and drinking more than usual. One night just before we opened the store, while driving home in his large Lincoln, he fell asleep at the wheel. He smashed into a tree about a mile from his East Shore Road house. He crawled all the way home from the accident and up the steep front steps.

The next day when I found out about the accident, I took him to the hospital. It turned out that he had two broken legs, a broken shoulder and a broken wrist. They operated all day and put most of his body in casts. He couldn't move, couldn't stand. I stayed at the hospital with him, and then Judy and I took him home.

About a week later, we were having a big grand opening party for the new store. Never one to miss a party, Will Rechler insisted on going. We got an ambulance to bring him to the store, then put him on the conveyor belt and sent him down the stairs. He was there in a wheelchair for the opening. On the day of the accident, he had met a new girlfriend, Joan Reccia Elkin, whom he would marry about a year and a half later, but more on that later.

Our new store was very successful. We ran a series of promotions that increased business, such as an annual George Washington Day Sale where Roger or Mitchell dressed up in a George Washington costume. Our normal gift and china and flatware business remained strong; we had pieces of jewelry that would sell as high as $75,000. We sold paintings in the

$15,000-$20,000 range . . . . Everything was going well. We didn't amass a fortune, but we were making money.

Next door to us was another high-end business, Hansen's Jewelry Store, run by two brothers. One day, in the middle of the afternoon, they were held up at gunpoint. One of the brothers resisted and was shot to death. After that, all the businesses in the area became more security conscious, including ours.

Our second year in the new location, we had an exceptionally good Christmas. On Christmas Eve, Roger and I were at the store until almost 11:00 P.M., and the whole shopping center was dark. We were in the back counting our money, intent on figuring out what we'd done for the month of December, and if we'd beat last year's figures.

Someone banged loudly on our front door, but we didn't answer it. The banging persisted. Afraid they were going to break the door down, I walked up front to see who was banging. A black fellow stood on the other side of the glass. "We're closed," I called to him through the door. "We can't help you."

He repeated over and over, "You've got to open the door. I have to buy the Leon in the window. You've got to open the door . . . ."

"What are you talking about?" I asked through the closed door, now thinking this guy was crazy. "We don't have a Leon in the window."

"Yes, you have a Leon that has candles." He explained that his name was Leon, and he was having his family over for Christmas dinner and wanted the Leon on his table.

I craned my neck to see what he was looking at in the window. I sure wasn't going to open that door. He was looking at "Noel" candlestick holders. All of a sudden, I realized the poor man must be dyslexic. Grinning, I climbed into the window to retrieve the candlestick holders, tossed them in a bag, and handed them out, wishing Leon a merry Christmas.

I remember a funny story from Mitch's childhood at Birchwood. I was in retail then and Gift Imports was open from 9:00 AM to 6:00 PM every day, and two days a week we were open until 10:00 at night. Judy would usually make dinner for me whenever I got home, no matter what hour it was. One night, I drove home late, tired and hungry, stopping the car in front of the garage. I jumped out to press the button to open the garage door, and when I got back in the car to drive inside my headlights were shining on the open garage. I couldn't believe my eyes! It looked like grass was growing on the floor of our garage.

I climbed out of the car again and walked inside. Sure enough, the garage floor was covered in sod. I looked around outside and discovered a couple of pieces of sod lying on bare dirt where my neighbor's lawn should be. Now, my neighbor was a nasty guy who never said hello, but the two pieces of sod were a definite clue that the sod should have been on his lawn, not in my garage.

Anxious to find out what happened and suspecting he might know, I woke up my son, Mitchell, age six or seven at the time. The neighbor had had his lawn sod that morning, and because they didn't like him, Mitch and a friend had lifted up all the sod and put it in our garage. I called my gardener who arrived at six o'clock in the morning to re-sod my neighbor's lawn.

When the boys did something wrong I made them a member of the "Five Finger Club." That would be one good swat on their rear end that would leave an imprint of five fingers. It got to the point where I hardly ever had to spank them. I'd only have to say: "Want to become a member of the Five Finger Club?"

Not long after the sod incident, another funny thing happened on my thirtieth birthday. When I asked Judy if she was planning anything for my birthday, she said no, so I decided to plan my own surprise party. Having thrown barbecues for 40 people there before, I invited 20 couples for a barbecue. Well, it turned out Judy was having her own surprise birthday party for me that same day. She invited five couples for a dinner party, telling them to dress in suits and dresses. I told my guests to dress casual. As I was talking to the men and Judy was talking to the women, husbands and wives were arguing about what they were supposed to wear. One of the overlapping couples was Roger and his wife, who figured out what was going on; two days before the party Roger told me. In the end, we had a party for 30 people who dressed however they pleased. Judy and I served an interesting variety of food and had a good time laughing about it.

One of the more nerve-wracking decisions of my life occurred in 1967 when I was thirty two years old. My father had been married to Joan Reccia Elkin for only two years then, and the new Gift Imports store was continuing to do well.

Dad was having a prostate procedure done and I accompanied him to the hospital, along with Walter Gross, his partner and friend. The doctor told us the procedure would take about two hours as they checked him in. Walter ran out to do something during the surgery and I sat in the waiting area, reading. After half an hour, the doctor returned, accompanied by another doctor and a nurse, with a paper for me to sign.

They had discovered my father had prostate cancer, and they wanted to operate.

"What happens if you operate?" I asked.

The doctor said, "He will be impotent, but the cancer will be gone for the rest of his life. You really have no choice, but to sign the paper." He was adamant.

I felt a great deal of pressure. Being only thirty two years old and I had to make this critical decision for my newly married father. . . . Walter wasn't there to consult and neither was Joan. After thinking it through, I decided this was simply too big a decision for me to make for him. I refused to sign the paper. I figured that if Dad wanted to have the operation, he could go in again for another surgery. Walter returned shortly and supported my decision, saying it was a good call.

My father came out of the anesthesia and the doctor told him he had cancer, explaining that while he had done the best he could, he didn't take everything out. I relayed the story to my father. "Good decision, son," he said. "I would have cut off your testicles."

In Gift Imports' eighth year in business, everything was going better than ever when a large Brooklyn department store, Fortunoff's, decided because of our success to open a Long Island location. They planned to imitate what we had done in an even larger format. For that whole year they had somebody standing in our store and following us around taking notes. There was nothing we could do to get rid of them.

That eighth Christmas we did our largest volume ever. However, when we looked at our annual profit picture we discovered we had made just $40,000. We'd had our most successful year ever and we'd hardly made any money. We'd invested an awful lot of work, time, and effort to make that little. We realized that prices would only get more competitive when Fortunoff's opened.

Around 1967 or 1968, we made the decision to get out of retail. I had enjoyed the store, learned a lot about marketing and merchandising, and gained confidence. There were long hours in retail and I wanted to spend more time with my family. It was time to do something else.

My father had already been very successful in real estate and was developing Vanderbilt Industrial Park (VIP) with Walter Gross and other partners. The way retailing was going, I felt there was more of a future in real estate. At that point, I had the confidence to give up the store and move on.

"Confidence is like a flashlight—no matter how dark it is you'll be able to find your way."—Donaldism.

My father wanted us to go into the real estate business on our own. Roger and I would manage the business while he would be a one third investment partner, inactive because he was already managing VIP. As a partner with Roger and me, Dad offered to buy a piece of land for us to get started on. It took almost a year to close the store because we had a huge inventory. We ran sales, cut prices, and sold off our merchandise—and after selling most of it, we held a final auction.

When I had originally told our landlord, Frank Castagna, that I wanted to leave he had been understandably upset. Gift Imports brought much of the traffic into the shopping center. "You have a 20-year lease, and you're only paying a dollar a square foot," he reminded me in an effort to get me to change my mind. I knew a dollar a square foot was an excellent rate then, and today it would be unheard of. The Americana Shopping Center draws the highest rent in the suburban market in the whole country today. Even then, Frank could get five to seven times what he charged me, but he didn't want to let me out of the lease because I was too important to the shopping center. He agreed to let us break the lease on one condition: "Only if you get me someone equal to you, someone who will bring in the same kind of traffic." That was a tall order.

I had inherited some of my father's luck, although I always considered mine to be "beginner's luck." What happened next was certainly lucky.

The day after I'd spoken with Frank was my Sunday off. That morning, I ate a leisurely breakfast, then picked up my *Sunday Times* and sat in my favorite chair to enjoy a second cup of coffee. Flipping through the pages of the thick paper, I spotted an ad for Brentano's, an upscale bookstore in Manhattan that also sold giftware, imported items, and art, such as lithographs and posters. "Boy, you can't get any closer than that!" I exclaimed aloud, nearly spilling my coffee. Hearing me talking to myself, Judy rushed in from the kitchen, and I told her my idea. Now, all I had to do was convince Brentano's.

First thing on Monday morning, I called Brentano's, and after introducing myself, I explained, "I'm a retailer with a store that sells product very similar to yours. It's in the wealthiest area of Long Island—the prestigious Americana Shopping Center. I think you could have a terrific store here. I could turn my lease over to you, or you could deal directly with the landlord."

"You know, it's very funny that you called, Mr. Rechler," he said. "We've just been talking about expanding to a suburban market, but we were thinking of Westchester first.

Tell you what, we have a board meeting scheduled for Thursday. Why don't you come in and propose this idea to the board."

I brought some of our merchandise to the board meeting. I showed them the figures: our volume per square foot, our gross sales, our inventory turns, and how many big ticket items we sold. They were very interested. Within a week, I put them in touch with Frank and they were working out a deal.

We held the final auction and sold out all our merchandise. The last day we were open, regular customers and people we knew stopped in to say goodbye. It was sad. Frank had been in the store many times, but that day was different—he came to say goodbye. He told me that if I went into business in real estate and ever needed a partner—or money—to call him. With that, he shook my hand and handed me an envelope. "This is for you."

It was still hectic in the store, so I thanked him and put the envelope in my inside jacket pocket. I thought he'd given me a little note or a card because he was that kind of person.

That evening, after everything was done I locked the front door for the final time. Gift Imports was officially closed. Taking off my jacket, I spotted Frank's envelope in my pocket. I sank into the chair at my desk in the back, also for the final time, and opened the envelope. Inside was a check for $60,000 and a note telling us that we did a terrific job and since we had acted as brokers he was paying us a finder's fee for bringing Brentano's to him. I was flabbergasted; we had just earned our first real estate fee! That, along with some money we had put aside for our new business, gave us our initial working capital.

Frank did well. He made three stores out of our space, which would become his formula as he upgraded The

Americana to the top suburban shopping center in the country. Frank eventually became a good friend and a partner in a real estate deal. He is Long Island's number one citizen, a classic example of giving back to the community. Brentano's did extremely well and was there for many years. It turned out like any really good deal should—a win, win, win situation.

Roger and I went on our way to start a real estate business that eventually would become the largest for industrial and office real estate in the Tri-State region. But it didn't happen overnight, and it wasn't without a lot of work both on our part and that of many others we worked with through the years.

CHAPTER 11

Reckson and AIP

My father was always the one to name things. As the third partner with Roger and me, he dubbed our company Reckson Associates, with the emphasis on son. While we were still closing the store, Roger and I were attending real estate and construction school three nights a week to learn how to read plans, do construction flow charts, and generally how to build. I'd had some experience with Pierce, Mayer & Grier in the sales end, but not too much in construction. In night school we met a friend, John Ruggiero, who would become my future tennis partner and Roger's lifelong friend. John planned to go into his father's masonry business, the largest on the island.

During that time, we drove around to look at land. In 1968, just as we were winding down our retail activities, we bought our first piece of land.

After Gift Imports was closed, we officially opened Reckson Associates as a full-time business, taking an office in Manhasset at 1295 Northern Boulevard, the same building where my father and Walter had their VIP offices. As he had with our first store, my father arranged for us to share an office with a lawyer friend in the building, an older Irish gentleman named Neil Mahr. My father likely did this so he could keep an eye on us.

Top: Builders and architects in training: Glenn, Gregg and Mark.

Center: Toody, Mark, Larry, Rick, Lenny, Glenn, Michael, David, Jeffrey, Steven, Mitchell and Stephen, our friend's children on the Fourth of July. Girls were a rarity, 1972.

Mitchell, Glenn and Mark in St. Thomas when Mitchell fell off the Sailfish.

top: A little 007 intrigue with Bob Yaffe on our trip to Greece, around 1977.

center: After our sauna escapade in Sweden we have a photo session with Bev and Judy, 1974.

bottom: The South of France, Eddie and I, around 1983.

Top: Mitchell's bar mitzvah, December 1972, L-R: Norman Rothstein, Stanley Rabinowitz, Bruce Tucker , me and Norman Berlin.

Bottom: Weekend in Boca 1982 with the guys in Peppertree, strictly domestic. L-R: Eddie Blumenfeld, Bob Yaffe, Richard Karyo and me.

top: Weekend in Boca with the guys: L-R: Richard, me, Bob and Eddie.

bottom: 1979, L-R: Norman Berlin, Stanley, me, Norman Rothstein and Eddie.

top: Camp Hi-Rock, the YMCA camp, L-R: Mitch, Glenn, me, Judy and Mark.

center: My boys and me celebrating our first day at the office at AIP.

bottom: In front of the Sphinx in Egypt, L-R: Me, Mark, Judy, Glenn and Mitchell. This was when the boys stayed out all night.

top: Mitchell, Glenn and Mark in London alone instead of in Greece.

bottom: My fiftieth birthday party on a boat, the Riveranda. The boys entertained all the time, now serenading me.

Top: Fourth of July party 1983, the Year of the Bulldog: Judy and me.

Bottom: Fourth of July party 1983, Cousin Ann, Robert and me.

top: Mitchell at the beginning of his career at Reckson with me in front of 395 South Service Road.

bottom: 1961 Ground breaking at VIP Hauppauge, before expressway. L-R: John Klein, town supervisor, Walter Gross, my father's partner, William Rechler, my father.

Top: Roger, Dad and I at the opening of AIP.

Center: My dad at the office.

Bottom: Roger and I at the opening of 225 Broadhollow Road.

top: Dix Hills, exterior rear view.

center: Dix Hills, family room

Top: Boca Raton, pool area, Bali-like.

Bottom: Boca Raton, great room.

top: Terrapin Hill, back.

bottom: Terrapin Hill, living room.

Family portrait at Mitchell's and Debbie's wedding in 1987. Rear, L-R: Jackie, Mark, me, Glenn. Front L-R: Judy, Debbie, Mitchell and Tracey.

bottom:Family portrait at Glenn's and Tracey's wedding. Rear L-R: Mark, Judy, Glenn, Tracey, me, Mitchell; front L-R: Jackie, Willi, Debbie holding Benji.

top: Family portrait at Mark's and Jackie's wedding. L-R: Tracey, Glenn, Jackie, Mark, Debbie, Mitchell; Judy and I are seated in front, 1991 at Terrapin.

bottom: Mark's and Jackie's wedding. L-R: Bennett and Bonnie Rechler, Tony and Amy Fromer, Jack and Ruth Wexler, Robert Fromer, Michal Fromer, Me, Ann Fromer, Gary and Nina Wexler, Yvette Fromer and Mitchell Rechler.

top: Me, surrounded by the ladies at Mitchell's and Debbie's wedding. Front row, L-R: Iris, Lilo, Bev, Cindy, Judy; back row L-R: Debby, Susan and Arlene.

center: Mark and family at his wedding, L-R: Todd, Scott, Gregg, Roger, Mark, me, Glenn and Mitchell.

bottom: Glenn and Mitchell having a fun time at Mark's and Jackie's wedding.

top: Twenty-fifth Anniversary party given by Roger at the Rainbow Room. L-R: Judy, Mitchell, me, Mark and Glenn.

center: My Sixty-fifth birthday party. Top L-R: Norman and Sandy Berlin, me, Arnie Gewirtz; Bottom L-R: Kenny and Roz Goldman, Roger and Evelyn Rechler and Clarc Gewirtz.

bottom: Standing at the bar at my sixty-fifth birthday party, L-R: Leonard Feinstein, Bob Yaffe, Paul Amoruso, Ed Blumenfeld.

top: Mark and Jackie's wedding. Uncle Jack and Aunt Ruth Wexler; Uncle Mort Rechler and me.

center: Glenn and Tracey's wedding L-R Stan and Iris, Judy and me, Peggy and Bruce

bottom: Windstar Cruise 1987 Pixhillians
Top L-R George and Helene Harasch, Judy and me
Bottom L-R Bob and Bev Yaffe, Bob and Madi Kaplan

top: L-R me, cousins Gary and Nina Wexler, Ann and Bob Fromer, Debbie (center), Gary and Shirley Friedman, Walter and Barbra Gross. Seated L-R Marty and Natalie Feldberg, Uncle Jack and Aunt Ruth Wexler. Surprise 39th anniversary at the Craft Museum.

bottom: At my sixty-fifth birthday party, L-R: Helene and George Barasch, Madi and Bob Kaplan, Natalie and Marty Feldberg, Shiela and George Leibowitz.

top: Twentieth Anniversary at the St. Regis Hotel.

center: Twenty-fifth Anniversary at The Rainbow Room.

bottom: We were married thirty years when Mitchell and Debbie got married.

top: Judy and I at Glenn's and Tracy's wedding at thirty-five years. Glenn was the last of the three boys to be married.

bottom: We had a Forty-fifth Anniversary celebration at home.

top: Thirty-ninth Anniversary party family picture, Top L-R: Glenn, Mark, Tracey, Todd, Lisa, Mitch, Gregg and Scott. Bottom L-R: Jackie, Michele, Debbie and Debby.

center: Thirty-ninth Anniversary surprise party. L-R: Arnie and Clara Gewirtz, Chuck Karst, Jane Weaver, Glenn, Debbie and Clark Weaver.

bottom: Mark & Jackie's wedding, 1991. Top row L-R: Norman and Sandy Berlin, Bob and Debbie Stillman, Peggy and Bruce Tucker. Bottom row L-R: Stan and Iris, Me and Judy.

top: Our Forty-fifth Anniversary. L-R: Judy with Helene Barasch, Susan Feinstein, Sandy Berlin, Claire Leinhardt and Madi Kaplan.

bottom: Surprise Thirty-ninth Anniversary party. Top L-R: Bruce Tucker, Stan and Iris Rabinowitz, Sandy and Norman Berlin, Bob Stillman; front L-R: Peggy, Judy and me, Debbie Stillman, Evelyn and Roger Rechler.

Top: OMNI - prize winning building called "The finest suburban office building in the United States."

Bottom: Naming of Reckson Center, Cradle of Aviation. Note from Tom Kelly, inventor of the LEM (module for moon walk) built by Grumman Corporation. L-R Mark, Gregg, Tom Kelly, me, Mitch and Todd

Dear Donald, From an engineer to a master builder, Regards, Tom Kelly

January/February 2000

real estate

portfolio

the source on reits & publicly traded real estate

Generations

Family Controlled Businesses

Rechler family makes cover of magazine, January/February 2000.

Savvy.
Charitable.
Definitely
Punch.

True to its name.

Donald Rechler, President of Reckson Associates Realty Corp., made the winning contribution to the prostate cancer research group CaP CURE at the "A Night to Remember" fund raiser sponsored by *Cigar Aficionado*

WINNING CaP CURE Contributor

top: Me at fifty.

center: I was on the cover of EAB's, 1989, annual report the year before they foreclosed on Omni.

below: Me at sixty-six in an ad in "Cigar Aficionado Magazine", donated at auction by Roger, Mitch, Gregg and Paul Amoruso at a charity smoker for CapCure at the Four Season Restaurant

Rhinos cheek to cheek in protective formation, safari Kenya, Africa 1986

center: Mark at Massai Village in Africa, always popular with the ladies.

bottom: Glenn and Mark serenade dancing hippos, along with David and Rick Yaffe.

top: Glenn famished after a balloon ride on the Serengeti.

bottom: The Yaffes and Rechlers waiting to leave on a safari in Africa.

top: First time all of us are on vacation together in Boca, 1986. L-R Mark, Jackie, Mitch and Debbie, Judy and me, and Tracey. Glenn took the picture.

bottom: First trip with the couples, cruise on the Wind Spirit, Christmas 1989. L-R: Mitchell and Debbie (pregnant with Willi), us, Mark and Jackie, Tracey and Glenn.

op: At Judy's fiftieth birthday celebration; 41% in the paper that weekend! L-R: Harvey, *rlene* and Richard Karyo, Susan and Eddie Blumenfeld, us, Debbie and Bob Stillman, Iris *nd* Stan Rabinowitz, Bruce and Peggy Tucker, Bev and Bob Yaffe, and Roger Rechler.

ottom: Judy's fiftieth birthday weekend, L-R: Arlene Karyo, Debbie Stillman, Susan *lumenfeld*, Judy and Bev Yaffe

top: Sunday drive in Chiang Mai, Thailand, 1989.

center: in front of the Royal Palace in Bangkok, L-R: Bob, me, Susan, Judy and Eddie.

bottom: Judy and I with Bruce and Peggy Tucker on a glacier near Jasper National Park, Canadian Rockies.

top: Karyos and Blumenfelds with us at Debby's and Scott's wedding in Sedona, Arizona.

bottom: The salmon I caught in the small boat in Gold Beach, Oregon 1993. The fish that keeps giving and growing story after story. Our trip with Madi and Bob Kaplan.

My Sixtieth Birthday party at the Hyatt Gainey Ranch in Arizona. Me with 4-1/2 pretty cowgirls: Tracey, Jackie, me, Judy, Debbie and Willi.

center: Me with my nieces and nephew L-R: Lisa and Gregg, Debby and Scott, m and Todd.

bottom: Four and a half mean hombres. L-R: Glenn, Mark, me Benji and Mitchell.

A hotshot divorce lawyer who was surprisingly philosophical, Neil Mahr was originally from a high-profile Manhattan law firm. He gave us an extra room where Roger and I put two desks together facing each other. We had our own office and the lawyer had his, and we shared the conference room and reception area. Roger and I hired a full-time secretary/receptionist. The older man enjoyed us, and loved to chat and give us life advice.

To start us off in business, my father bought Blydenberg, a five-acre property in the Hauppauge part of Islip, just off the Expressway. A very wild piece of land, it started at about 125 feet wide in the front and then ran from one block to another, 100 feet deep, the shape of an arrow. From block to block there was about a 600-foot elevation drop. Because of the grade the only way you could build was with interior lots—to have one building behind the other, and only put two buildings on the street. In order to build what made sense—four buildings, two on interior lots—we needed five variances and two roads abandoned. This would be our first construction project.

Every obstacle you could have in construction existed at Blydenberg. We called our first construction project, *trial by fire*. By talking to almost everyone and the planning boards and building department, over and over again, we learned how to create rentable industrial space out of the mess it was. I don't think our father picked it for that reason, although you never really knew with him, but it became a great learning experience for Roger and I, and we eventually succeeded in making it work.

Our first two buildings were multi-tenants with loading in the front. With all the problems we had to overcome, the first building cost us six dollars a foot to build. My father was upset because he thought it should have come in at four dollars a foot. We made the units as small as 2,000 feet, and renting to the smaller tenants was always tough because the negotiations would come down to literally pennies. Our first tenant was a 2,000-foot ceramics school, and I recall our first rent was something like 56 cents a square foot.

It took us two years to complete the project, from wading through the variances to renting the last space. Considering all the problems, we did a good job. As industrial projects went, this one was still on a fairly low scale and therefore a good beginning for us. When I had marketed and run the retail store I didn't have to worry about having the lowest prices because we had the best quality merchandise. Here we had to have a low rent to compare with other choices, because our product wasn't any better than other buildings.

Shortly after opening the new office, I went to an auction and bought Allwood, a 20-acre piece of industrial land in central Islip. It was another low-end deal like our first one, but with larger industrial buildings on inexpensive land that we bought for $12,000 an acre. We wanted to divide the property into lots, lease each to an end user for his business, whether manufacturing or warehousing, and build the plant he needed. It was successful, but again very price oriented because it was a mid-to-low-end product.

We had some success, but word on the Long Island real estate scene was that it was because of our father. That never bothered me because I was very comfortable being Will Rechler's son.

"No one sees things as they are but sees things as we are."   *—Donaldism.*

As I said before, my father lived through the Depression, so he was a very cautious deal-maker. In order to minimize risk, he looked for the least expensive property and always had lots of partners, people he knew.

A friend of Dad's, Jack Friedman, along with his father-in-law, owned 88 acres on Veterans Highway in Islip, near MacArthur Airport. In 1971, when we looked at this acreage as we were nearing completion of Allwood, the only planes landing in MacArthur Airport were small, privately-owned planes.

Ever cautious, my father thought it would be a good idea if we became a one third partner, rather than a half partner. He negotiated a third for the three of us, while Jack Friedman would retain a third, and Tom Manno, a partner from VIP, would take a third. That meant Roger and I would each have only 11% of the deal, and we would be the working partners. Nonetheless, Dad put the deal together for us. Friedman wouldn't agree to it until he interviewed Roger and I separately, but he was duly impressed, and we moved ahead with what was to be our third deal.

We were excited. The property had tremendous potential; it was 88 acres directly across from the airport and a beautiful wide highway, Veterans Highway, ran along the front. Even though it was way out east—a lot farther out than any other industrial parks—I recognized it had a lot of good selling features. I was excited to build on it and pull business out there.

This was not easily done. At that time, the farthest east that people were going was the Hauppauge area, where my father had built his park, between exit 54 and 55 on the Expressway. It was hard to pull anyone past exit 56, and our new property was at exit 57 and South. We needed a very good, marketable reason to get them all the way out there.

We built our first buildings. This time, we specified that we didn't want any garage doors showing and everything had to be brick on three sides. We gave the buildings an upscale look, hoping to draw upscale manufacturers that argued about nickels and dimes instead of pennies, and who would allow us to make a decent profit. From my retail experience I knew that you were able to make a higher profit on better merchandise.

After our major learning curve with Blydenberg, we now worked our way through all the problems more easily and jumped through the necessary hoops that would allow us to do what we envisioned. We filed a map that would allow us to have flexible lot lines and put a road system in that would maximize the property and bring water and electric to the property. We convinced the town of Islip that we were going to give them a model industrial park that would be even nicer than Smithtown, which my father had built in Hauppauge. This one would really be something special. They went along with our ideas and it all came together at a fairly good pace. We got our map approved and our roads in and approved. The park was called Airport Industrial Park, and we carried on the airport theme by naming our roads Orville and Wilber.

Initially, we built a spec building—a fine-looking brick building, about 32,000 square feet that could be divided in half. Our first tenant was SITA, a Belgian computer company. They handled all the air traffic computers in the area, and this would be their new headquarters. SITA needed a huge computer room. The standard was to allow 10% for office finishing—a finished area in an industrial building—because otherwise banks wouldn't lend on it. SITA wanted almost 75% of the building finished with air conditioning, including their huge computer room, but they didn't want to pay for it. They wanted it included in the rent.

I had some understanding of the mortgage business from my days at Pierce, Mayer & Grier, and in the meantime, had become good friends with our mortgage broker, Ken Goldman. Ken introduced me to the head of the real estate lending department at the Dime Savings Bank, one of the largest independent banks at that time. Over lunch, Ken and I worked on selling him on the property and the plan. I tried to convince him that even if we finished the building 80% and the tenant wanted out, there would always be another to take their place. Long Island was an R & D (research and development) hub, along with Boston and San Francisco, and there was continual hi-tech on the island.

Many of the big companies that had made airplanes for World War II were located on Long Island: Grumman, Fairchild, and Republic. Already a technical base, it was only going to expand. Island workers had a good reputation because women on Long Island had been the first to take second jobs during the war years. There was even a famous poster showing them working in the airplane factories with the slogan: "Uncle Sam Needs You." The poster girl was called Rosie the Riveter. It was a great work ethic to market.

With the banker's support, we were the first to build totally finished one-story industrial buildings with hi-tech rooms, and then rent them to research and development companies. After SITA, we rented to a Japanese importer. Then we signed a lease with Aimes, a French company that made jet engines and wanted to be next to the airport. (We later found out they would assemble French engines for the Mirage aircraft and send them from the U.S. to Israel, thus keeping Russia from discovering that the French supplied the Israelis with aircraft.) Before we knew it, we had four foreign companies in the park. Shortly after our park became successful, the area became a free trade zone and attracted even more international companies.

Around this time, my father reached back to his old friend, Murray Groberman from Porch & Patio, who now owned a bowling alley in Rockville Center. We weren't all that successful when we'd shared a store with Murray, so we should have learned from that, but he was selling his building and Dad wanted to buy it and convert it to office space. It was a two-story bowling alley, supposedly 52,000 feet, a split level with spaces that were difficult to figure out. Sunrise Highway seemed liked a good location, but I didn't know much about it. Another one of our early trial-by-fire projects, it was more like *trials and tribulations*.

We bought the building and things went downhill almost immediately. On the very day we closed and signed the papers, there was an unexpected freeze that night and the pipes and sprinkler system froze. We had all these beautiful hardwood floors from the bowling alley that we thought we were going to be able to use for the offices along with big plate glass windows. The floors froze and rose; they were totally ruined.

By the second night we hired a guard because we had to have somebody watching the place. There was no heat, and when the guard came in he put a faulty electric heater in the little room where he sat. It caught fire. Most of the inside of the place was burned out by the second night we owned it. What windows, doors, and electrical wiring that survived the floods and the fire, the firemen destroyed.

About two months later, we were told that the Long Island Railroad running next to the property was going to expropriate part of the parking lot. They took half the land and our amount of parking space was cut in half. As if the fire and flood hadn't been enough, now we had a mess on our hands that just kept getting worse. It was Murphy's Law building: "Anything that can go wrong will go wrong."

As we were cleaning up, we tried to figure out how we were going to divide the tenant spaces up and finally brought in an engineer to measure the place. It turned out we never bought 52,000 feet, and the whole building, with all the split levels, was closer to 40,000 feet. We'd been lied to about the size and had overlooked the fact that the number wasn't in the contract, so in addition to all the other problems, we had overpaid by 25%. I'd never felt right about the deal, and this experience made me realize I should pay more attention to my instincts in the future.

In about four weeks we had gone through more plagues than Moses brought on Egypt. Finally we got it all cleaned up. However, we realized that because of the nature of the space we weren't going to be able to rent offices. Rethinking our plan, we wondered if we could attract the high end R & D tenants like we had out at the airport, but at an even higher price because it was such a westerly location.

It wasn't an unfinished warehouse because as a bowling alley it had been a finished air-conditioned space. We attracted a medical lab that did blood work and other tests, but they had only been in business for less than a year. They wanted to take about 15,000 feet, which we had to outfit. It was a very expensive installation because they needed extra plumbing and other special things. But we were anxious for tenants and gave them a six-month concession to come in rent free. At the end of the concession period they went bankrupt. This experience was a tough lesson from start to finish. It took us another two years to finally get a tenant that made Rockville Center a viable property. The tenant in the garment industry was from South America, and eventually bought the building from us.

With the success of Airport Industrial Park, our company continued to grow and we began working with brokers. There were very few office buildings on Long Island in 1973 and industrial brokers usually made the office deals. Island Realty had two office brokers, Mel Loberfeld, and a young, good-looking man who worked with him, Sam Rozzi. Mel was one of the oldest brokers on Long Island, having worked in New York City prior. Sam had a nice demeanor and still is active in the industry today.

We now began to get tenants at a very good pace because we were the only ones able to do all their finishing and add the costs into the rent rather than asking the tenants for payment up front. That formula allowed them the capital to keep working and later became key to our success. It finally allowed us to market a high quality building at market or below market and have a unique product.

Mel came up with an idea. "Met Life is looking to move some of its offices out of Manhattan," he told me. "They've got to be 50 miles away from Manhattan because they don't want any trouble with the unions. If they're closer than 50 miles, they have to have the same union and they really want to get away from it."

Airport Industrial Park was exactly 51 miles out of Manhattan. Mel tried to sell them on it, but they said they were not looking for an industrial park. I don't remember how it happened—either my father came up with the idea or I did—but we quickly changed Airport Industrial Park into Airport International Plaza (AIP).

We started going into Manhattan to meet with Met Life. The first thing we had to do before we could even begin to negotiate was convince them AIP could be viable as an *office* area. They didn't believe they could get qualified staff to go out there, and the only way I could change their minds was to prove them wrong.

We rented a trailer, had it delivered to the site, and set it up as a makeshift office for their human resources people. We even paid for their personnel ads and kept our fingers crossed hoping future employees would show up for an interview. The morning of the interviews, Roger and I drove into work at eight o'clock, anxious to see if the ad generated any response. We couldn't believe our eyes. The 500-foot long block was lined up with so many people the human resources team couldn't possibly handle them all. Within the next two days, they attempted to process between 200-300 applicants.

This level of negotiation was new for us—office space with tenants of this caliber and size. Met Life was difficult and took advantage of our inexperience and enthusiasm to be successful. We offered them extra built-in expansion, extra parking, and everything else but our blood. Finally, we settled on a price and drew up the lease agreement. But on the day we went in for a signature on the lease they said they would only sign the lease if we gave them 50 cents a foot less. I just sat there devastated. They were renting about 24,000 feet of a

36,000-foot building. This was big for us. I knew that once I had them in the park, it would raise us to a new level, so they had me over the proverbial barrel. I needed them, even at a loss, and they knew that.

At the time I was in my early to mid-thirties and Roger was in his late twenties. Mel who was brokering the deal was in his mid-fifties. Mel slammed his attaché shut with a bang, stood up, glanced at me, then said to them, "I'm not letting you rake these guys anymore. Let's go, Don."

I had no choice but to follow Mel out of the office. "Are you sure you're doing the right thing?" I nervously asked him as we were leaving.

"You can't let them hold you up. Word will get around," he said.

Three days later after holding our breath, we got the lease at the original price. I learned a valuable lesson that day. Sometimes you've got to say no and walk away. There's a point you have to walk away even though your tongue is hanging out for the deal. It won't always go your way, but more times than not it will.

A funny incident happened just after we rented Met Life and were dividing (putting offices in) and finishing the building. We had a guard there all the time. One day, Roger and I noticed footprints through the newly poured cement walk, so we had to have it redone. The next day, we saw this same guard leaving by the front door and stomping through the wet cement like Inspector Clouseau. That alone should have been a strong clue for us to get rid of him, but it was hard to get help then. Shoreham, the nuclear plant they tried to open on Long Island, was being built then (1974) and they were hiring all the good contractors, electricians and other trades people; you couldn't even get a decent guard from the security companies.

Finally, we got all the windows in and everything completed, and Met Life was scheduled to move into their new offices the next week. We got a call from the security company to say that people were shooting from behind the building and all the windows had been shot out. I called the police, then jumped in the car and drove to AIP. The police were already there and I walked around with them. Stopping to look at one of the windows, I said, "Why is the glass inside? If somebody shot the windows out from the outside wouldn't the glass fall out?"

The cop agreed. Upon questioning the guard, they uncovered that, having nothing to do, he shot out our windows so he could look like a hero fighting off the villains. The guard was charged with vandalism.

We replaced the glass and Met Life moved in without further mishap. It had been yet another learning experience for us that would continue with our new tenants, although in a very positive way. One of the things we learned from Met Life was that not only did they pay their AIP people about 25% less than their Manhattan employees, but we heard they got about a third more production out of them. They had a report with this information, and although we tried to get our hands on it, they didn't want to give it to us. However, we used everything we learned, and started to rent more office space.

In 1971, New York State awarded us "Best Industrial Park." Airport International Plaza was the first hi-tech park in the state, so it was a model. In an effort to attract more of this type of business, the state Economic Development flew in international people and potential customers from other states directly into our park by helicopter. SITA, a Belgian company, Aimes, a French company, and a Japanese importer had all come to AIP from this initiative.

CHAPTER 12

Life in Dix Hills

Meanwhile on the home front, our third son, Mark, was born on May 17, 1966. With our growing family, the Jericho house was now getting too small, so Judy and I talked about buying a larger house. In 1968, when we started Reckson Associates and began to look for our first piece of land, I was also thinking about building a house. On my way out east to look at industrial building sites, I regularly drove past some very good looking contemporary model homes, off the expressway between exit 51 and 52. One day, I wandered in and they told me about a new section being built on Astro Place, off exit 50. At that time, they only had one house sold in that section.

I quickly took Judy back out to see the models. She loved the houses. I took my father to see the lot I was considering and he hated it because it was totally flat. He was right. I took Roger out and the two of us walked up and down hills, through mud and snow, looking at every lot. We picked out a wooded lot with a big hill. It was farmland, but that particular lot had windbreakers (trees that stop the wind) and a sump to catch the water so it wouldn't ruin the farmland.

We were able to lay a 100-foot ranch style house out beautifully on top of the hill. Later I was able to landscape the whole back piece of the property, so that it looked like two or three acres instead of one acre because of the shape of the lot and the topography. Three or four years down the road, after I started to become very successful, we put in a swimming pool and a tennis court.

In 1968, the house cost $42,000 to purchase and build. I made eighty-four changes, which included $22,000 worth of extras because I combined two models and added air conditioning and special lighting for our young but decent sized art collection we had started to develop at that time.

Dix Hills was a perfect neighborhood to bring up children. The neighborhood was mixed, in a very good school district and the schools and high schools were large. Very important to us, our children were able to meet a large and diverse group of other kids, which was happening less and less on Long Island.

We moved into the house in December of 1968 and it was too cold to put in the driveway up the big hill, so we had to wait until spring. We lived in a muddy, noisy construction zone that first year, our house being only the second one finished in the whole development.

My cousin, Ann, had married Bob Fromer. As our first visitors, Bob and Ann Fromer drove up the rough slope and parked in front of our new house. A trim, good-looking family man who always seemed to look the same as the first day I met him, Bob was a partner in a law firm. He was not only a bright corporate attorney but a sharp, street-smart businessman which served him well. Ann and I had always had a special relationship, having grown up together, and the couple would become our close friends. They had been looking for a house in Great Neck, and Bob had just turned one down because he felt they were asking too much money. After they tromped through the mud and came inside, Bob asked if we had a phone. He called the lady he'd just turned down, told her he'd changed his mind, and arranged to buy the house in Great Neck. Seeing how we were living, he suddenly felt the Great Neck home was a great buy after all.

The year after we moved into our new house in Dix Hills, Bob Yaffe, a real estate broker I knew from the industry, bought a place two blocks from us. I invited Bob and his wife, Bev, to come over to the house to meet Judy, and the girls hit it off immediately. Judy and Bev became the closest of friends, and so did Bob and I. A total gentleman, equally comfortable in the role of a good friend or as a long-time business partner, Bob had as mild a demeanor as Clark Kent, but the resolve of the "man of steel." Bob was detailed and meticulous in everything he did, whether it was real estate or planning a trip. As every friend's photographer, whether Bob was shooting a party or trip, you'd get a professional photo album the next day.

Like us, the Yaffes had three boys, Allan, Rick, and David, all friends with our boys. David became Glenn's best friend as the two were very similar: both boys were bright and creative, and neither liked to listen to authority. Their teachers were constantly calling conferences with their mothers.

In 1976, Bob and Bev got us signed up on a trip to Israel with the group that was building the new temple. It was a wonderful trip with a good group of people and we made some new friends. That was when we met Richard and Arlene Karyo, who would become very close friends.

Richard Karyo had a different philosophy on life than any other of my friends. He had a terrific head for business, with a long-term perspective, which was different from most of us. Richard was born in France. His family were prominent department store owners in Paris. His father, André was a Holocaust survivor. Richard and the rest of his family hid in the countryside, outside of Paris, during the war. After the war, they fled to the United States, when Richard was nine. His parents sent him to summer camp immediately upon arrival, speaking only French. "Frenchie" met Arlene (his bride to be) at camp when he was twelve. Soon after, you couldn't tell that Richard was French since he had no accent at all, but the accent was the only part of France that Richard left behind.

On this trip we also met Helene and George Barasch, the Bolnicks, and Shotskys. All were from Dix Hills, along with Madi and Bob Kaplan, and today all have homes in Boca West where we still remain friends. We call the group the "Dix Hillians."

On the plane to Israel, we won a prize for having the lightest luggage going. On the return trip, we had the heaviest luggage because of buying so much art and antiquities. On all of my trips, because of my experience as a buyer for the retail store, I had no trouble buying the most in the shortest time. I had it down to a science. From that trip on, I was dubbed "Meyer the Buyer."

We traveled all over Israel, saw all the sights, and along the way ate in many different restaurants. After exploring the tunnels where King David's old temple used to be, we went to the Wall and later the Red Sea. We journeyed to the borders with Lebanon and ended up all the way south in Eilat, which I remember was terribly hot.

Our tour guide was an Israeli tank company first sergeant who spoke to us as if we were soldiers under his command. I just mentioned that Glenn didn't like listening to authority figures—well, the apple doesn't drop too far from the tree. I don't take kindly to authority either. If the tour guide told us to march left, I went right. If he wanted us to stop and listen to him I walked over and read the information on the wall.

The guide was getting frustrated but I really didn't give a damn. In fact, I noticed three others behaving the same way: Bob Yaffe, George Barasch, and Rich Karyo. Up until then, I hadn't moved too far past my college friends, except for Bob. These three were the newest of today's now *old* friends, and I can see why we bonded during that trip. None of us wanted anything to do with following that guide like sheep. They all had a great sense of humor, and good insight into people and life.

Visiting Masada was the most inspiring part of the trip. Masada was where the Israelites fought off the Roman Legions for many years until finally they knew they couldn't hold them off anymore because food and supplies were running out. Rather than become slaves of the Romans, the Israelites killed themselves. To the defenders of Masada, freedom was more precious than life itself.

At our Dix Hills home, even before we put in the swimming pool and built a proper tennis court, we built a "paradise tennis court" on the lower level (four ping-pong tables together). It was there that we started our famous Fourth of July barbecue, a tradition that would go on for many years.

As we snapped the yearly Fourth of July pictures we started to realize that almost all of our close friends were like us. They had all boys, except for the Rabinowitzes who had two boys and one girl. The Tuckers had four boys, and finally a daughter. The Kaplans, whom we were just starting to know then, had three boys. The Yaffes had three boys. Sandy and Norman Berlin, Judy's sister and brother-in-law, had two boys, our nephews, Stephen and Lenny.

Early on, our Fourth of July parties were like a boys' camp. When the boys all started to go away to camp in the summers, we changed our traditional party date to Labor Day. Sometimes we had adult parties, but in the early years we had parties with all the children. As the boys began graduating from college, Judy and I restarted our Fourth of July family party tradition. Over the years, with the children or without them, we probably had about 35 Fourth of July parties. For every one of them, Judy would make the family peach cake and I would make my special Sangria by the gallons and we never seemed to have enough. (Recipes included.)

In 2000, we finally gave up having the parties because it became too much for us. By then we were inviting close to 300 people: our friends and their children, all my boys' friends and their children, along with some of their parents. Everyone invited would come, no matter what the weather. Every year, whatever date we chose, it might be pouring in the morning, but if we were to serve lunch or dinner at say one o'clock you can be sure that by then the rain would stop; it even happened during the worst storms when it was projected to rain through the whole day. Everyone joked that I did a rain dance to make the sun come out.

Mark was about four years old when we built our pool. One afternoon, my father and Joan were visiting and all of us were sitting around the pool. Mitchell was in swimming and Mark had water wings on. All of a sudden, when Mitchell screamed we discovered that Mark had disappeared and no one had noticed. Mark had taken his water wings off, thinking he could swim and went straight to the bottom. Somehow he was able to crawl along the bottom until he caught Mitchell's feet and his brother pulled him up. From that day on, we made sure all the boys could swim well.

I had shared my mother's love of dogs growing up, and now I shared my love of dogs with my sons. We had two little poodles in Dix Hills, although one eventually got run over in the driveway. We replaced him with a bulldog, named Winnie, the second of three bulldogs we had over the years. Whenever the kids or I swam in the pool, Winnie ran back and forth barking, then he would stand at the deep end barking at you as you swam a lap. One day, Winnie fell in. All dogs are supposed to be natural swimmers, even my other bulldogs swam, but Winnie went right to the bottom of the deep end. I had to pull him out, and fortunately, in the end, he was fine. But have you ever tried giving a bulldog mouth-to-mouth resuscitation?

When we moved to Dix Hills, we joined a Reform Temple in the town of Huntington that was almost 25 minutes away. Occasionally, I drove the car pool. The kids loved me to do it because they used to get me lost. A couple of times I didn't get them to the temple at all, and often we were late because they misdirected me on purpose.

We lived too far away from this temple to make any good friends there, and the temple in our neighborhood was too conservative for us. Another closer temple near us, where most of our friends belonged, used to meet in a house and then later in a Long Island development center. I didn't want to get involved because I knew eventually they would start to build a new temple and I remembered how involved my father had been when they built the new temple in Whitestone. I was just too busy for that.

A couple of years after we moved to Dix Hills, we took the boys on an island-hopping vacation in the Caribbean. Mark was four by then, Glenn was eight, Mitch was about 11, and we figured the boys were old enough to travel. We went to Puerto Rico, then to St. Thomas, and finally to Jamaica—not quite the same route as our honeymoon.

After flying to Puerto Rico, we took smaller planes to the other islands, the kind where you have to go out on the runway and climb the stairs to board. Several times we were late and had to run across the tarmac to catch the plane. One day, Mark looked up at us and said, "I like vacations, but I get so tired from all the running on them."

Just before we were supposed to go to the airport in St. Thomas, I took Mitchell out sailing in a sailfish. As I was tacking in to the shore, I saw a shadow in the water. I came about to avoid it. The shadow followed. It was a shark or some other large creature in the water, chasing our boat. As I tried to outrun it, Mitch fell off! I quickly grabbed his ankle and yanked him back aboard. Mitchell was terrific the whole time.

Finally, I sailed in and we landed safely. By that time it was getting late and we were due at the airport. Back at the hotel, Judy took one look at me, "What happened to you?"

"I'll tell you later," I said, brushing past her, realizing we might miss our plane. "Just give me my clothes, so I can change. We have to go." I pulled on a pair of pants, then grabbed the luggage. Mark and Glenn were ready, but at the last minute we couldn't find Mitchell. Outside, we heard sobbing and found him in some bushes. He was understandably upset from our harrowing experience, but he didn't want me to see him cry. Judy comforted him, quickly got him ready, and we made the flight without a minute to spare.

The Caribbean adventure was the first of many eventful family vacations. Shortly after we returned, Mark decided to take his own little vacation and ran away from home. He left us a crayon drawing note, and got as far down the driveway as the back of his mother's car.

Unlike most of the other parents in the area, we never sent our boys away to camp for the entire summer. They only went away to four-week camps, so we could spend the rest of the summer as a family. We chose a YMCA camp in Connecticut, thinking that because the YMCA took a lot of kids on scholarship who couldn't afford to go to camp, the experience would provide our boys with a better feel for a diversity of people and that was important to us. Mitchell attended for three summers, Glenn for two, and Mark for only one. After that, we started to send them to two-week specialty sport camps.

Mitch went to tennis camp in Manitoba, Canada with a renowned tennis teacher. He came back a vegetarian because his teacher was one. His mother cured him of that in about ten days by buying a how-to book on vegetarian food and cooking it for him. Finally, with all the beans practically coming out of his ears, he finally wanted a hamburger.

From the time Mitchell was three years old he had such a beautiful golf swing that people at driving ranges stopped to watch him. Some thought he looked like a pro in the making, but the truth of the matter was that he wasn't hitting the ball that well. Mitchell always enjoyed golf and did have great form. After he went to golf camp, his game improved.

When Mitchell was in the sixth grade, he went to New York City to play in a chess tournament and won the sixth grade championship for the entire area. When he came home, he taught his best friend, Hal Skopicki, who lived across the street, to play chess, and after about two days, Hal beat him. Hal was an exceptionally bright boy, and an excellent athlete with tremendous determination. Today, he is the head of cardiology at North Shore Hospital, the largest hospital on Long Island. The same thing happened when Mitchell came home from camp and taught his best friend how to play golf. The second time out, Hal beat him. Mitch went downstairs and threw his clubs away. I retrieved them and put them in a corner, but he refused to touch them for years. Today, golf is his favorite sport and he is a good golfer.

Living in suburbia, the boys needed bikes to get around, even to go to a friend's house. Mark had lost one two-wheeler already and I'd warned him that if he lost another bike it would be the last one he would ever get. He often rode over to visit his friend, Robbie, who lived on a main road, and would leave his bike there, so I think that was how he lost his first one: someone helped themselves to it. We surprised him with a new bike for his ninth birthday and Mark excitedly rode it around the neighborhood.

The next Sunday morning when I got up and went to the kitchen, I noticed someone had left the refrigerator open and milk on the counter. I put the milk back, closed the refrigerator, and hollered at Mark. He said, "Dad, I didn't have milk."

"One of you must have had it last night," I told him. "You're going to ruin all the food, leaving the door open like that. It looks like everything thawed out."

Mark went downstairs to get his new bike to go for a ride. From the kitchen I heard him scream that it was missing from our basement. He ran upstairs and I hollered at him again.

"You left it at Robbie's." He insisted that he didn't. I told him to go to Robbie's house and check.

When he came back, he was hysterical. The bike was nowhere to be found. "Well, that's it," I said firmly. "You're not getting any more bicycles." Mark walked away, sulking.

About one o'clock that afternoon, I went into our playroom to watch the football game. (We had put our cars under the house and turned the garage into a finished playroom with a pit in it, at the time a popular thing to do.) I stood in the doorway in shocked surprise—the television was gone. Looking around the room, I could see that we were missing a few items. It turned out that we had been robbed of everything valuable on that side of the house, from the kitchen on back. Winnie, the bulldog, slept on the couch in the den. He was a big, mean-looking dog with a loud snore, but absolutely didn't hear or see anything well. However, the robber must have been afraid to walk past him.

We reported the robbery and the police thought they knew who did it, but they couldn't prove it. They found our TV hidden in some bushes outside where it had been sitting all night because the robber apparently had a small car and couldn't fit the TV into it. Mark's bike was never recovered. I bought him a new bike and apologized to him.

When Mark was a little boy, he was a collector. He collected everything that belonged to his brothers and me. Mark had a fascination with taking things apart and liked to tear apart his brothers' things. They must have considered him quite a pain-in-the-butt little brother. I had a stamp collection that I'd had since I was a boy and many of the stamps were mint blocks and mint sheets. One day, when Mark was about five, he climbed up on the top of my closet, took my stamp collection down, and tore all the stamps apart. Needless to say, I was upset.

When I came home from work I drove up the driveway at Dix Hills and into the garage. For some unknown reason there were often a couple of cement blocks in the way. I would have to get out and move the blocks. One day, I returned home late in the afternoon to get something out of my bedroom. I thought I heard voices, so I went out on the balcony and looked down between the slats. There was Glenn, then 12, standing on those cement blocks kissing a tall girl! They would neck under the balcony by the garage so no one would see them.

The boys started being popular with the girls young—Mark even younger—and the phone never stopped ringing. It got to a point that I wouldn't answer the phone if I was the only one in the house because Mark would get calls from literally fifteen to twenty girls with names that I could never remember. All I would say to Mark was, "Some French girl called you" because to me, the names were foreign when they were Denise or Mona.

In those years, I was very busy starting the real estate business. Mitchell had started Little League by then and I caught as many games as I could take time off to see. Even though he was a good ballplayer, he went through a year where no matter how much I worked with him, he couldn't hit the ball. He managed to get on base often. He was a sensational fielder and he played third base, but he went the whole season without getting a hit. At the same time, it never affected him and he went the whole season without making an error. At the Little League awards dinner that I attended with him that year, they presented him with a special award for determination and being the best fielder in the league.

When Glenn was about fifteen or sixteen, he went to a lot of parties. One night when Judy was away, I waited for him to come home. At about one in the morning, I finally went to sleep. A couple of times while I was sleeping I thought I heard the phone ring, but I was too tired and it didn't make sense to me. About two o'clock, Glenn came into my bedroom and woke me up. He was dressed for bed, although he hadn't been to bed yet. He was shaking as he told me that he'd just heard from his friends who were being taken to the hospital to get their stomachs pumped because they took some pills that had bad stuff in them that they weren't aware of.

I had been having some allergy problems and had started on a vitamin regime. Having seen the doctor do miracle things with vitamin C, I stirred up a twenty-ounce double dose of vitamin C and made Glenn drink it down. Shortly after, he started to calm down and relax. The next day, I calmly talked to him, making sure he understood how careful he had to be while his friends put pressure on him to experiment with substance. I think at that point our relationship strengthened.

CHAPTER 13

Appointment
to Say Good-bye

Airport International Plaza won the state's Best Industrial Park award for three years running. We also won a landscaping award–both for New York State and nationally for the Best Landscaped Industrial Park. We were very proud of our awards, and in addition to the recognition, the park was growing financially successful. By this time we had built a very loyal following. Some of our contractors had previously worked for my father, and many of the sub-contractors had been with us for years and years. DeFazio Electric, Ruggiero Brick Masons, Maybet Carpenters, and Apollo HVAC are just a few, and we used Dad's accountant, Gary Friedman. Many people were with us for over 30 years, and as I write this, we still use them today.

By 1972, we had completely transformed our vision for AIP into reality. After our first development, our goal as I mentioned, was to produce high quality product at market or below market and give customer satisfaction, in essence, to repeat our retail strategy in real estate. In retailing the customer was always right. At AIP we were finally able to practice what we believed: "If you do something, do it with class; if you create something, create it with quality; if you start something, finish it with enthusiasm."

We came up with a formula at AIP, that we continued to use in all our future projects: we gave extra land and more parking than anyone else did. Our reasoning was that this would allow us to add more office space in the future, as we foresaw the hi-tech explosion. With a higher quality building we could attract those tenants and get higher rents as more and more hi-tech firms began to look for research and development or office space. We worked with good bankers, and had built a reputation of trust—of keeping our word to town officials, brokers and tenants. The one problem that remained as we drew to a successful conclusion of the first section: despite all our effort and success as the working partners, Roger and I only had 11% each of the income of the park.

AIP started to enter its pinnacle in 1973 when, after doing the Met Life job, we gained enough confidence to build a unique fifth building. It was the first one-story atrium office building on Long Island, a building with a garden in the middle that featured interior offices with windows. We were able to break it down into lots of small offices or R & D because it allowed for more window space. The unique atrium offices were met with enthusiasm and brought in higher rents.

In 1973, we moved our company offices from Manhasset to the new atrium building in Bohemia. My father, pretty much retired by then, also had an office there. It was a place to hang his hat when he came out, because he liked to keep his hand in the business and still showed up at the VIP offices in Manhasset occasionally.

To promote our new atrium office building, we threw a major party for all the brokers in the industry and state and town officials. There must have been more than 250 people at that party. We had helicopters flying over the whole area, giving people rides, and we brought in a lot of antique cars for people to ride around in, and the food was excellent. From that point on, our parties became the hottest ticket in town. Wherever we went—New York, Westchester, Connecticut—our reputation preceded us. Every time we gave a party we drew the full real estate community, whereas our competitors didn't get that kind of draw.

We finished off the first section of AIP by adding Periphonics, another hi-tech firm that later became a public company. They invented the first voice answering system with computer, the recorded voice on the telephone that is commonly heard today and is so annoying. The last lots in the section were leased to an Italian company, Castelli Furniture, and Merganthaler, a Fortune 500 company that made printing presses.

By 1974, we were ready to develop the second 86 acres at AIP. This land was still owned by Ebin, the European fellow who had originally sold Jack Friedman his land. For the past year I had been negotiating for it, but couldn't get the price lower, even though the land didn't have any frontage. Ebin finally gave us a PM mortgage, which means that over time we paid for the land as it was released for us to build on, until we would eventually own the whole piece. As expected, we ended up paying 125% of the price of the land as each lot was released, then the last 25% was released free.

We had moved through the first section very fast and the partners Jack Friedman and Tom Manno were happy, but Roger, my father and I still had the smallest percentage of it. Since this was a new section I suggested that we go back to our partners, Jack and Tom, with a new deal. We wanted to offer them 25% each on the new section instead of a third, so we could each have a larger share. This would raise Roger and I each up to a 17% share. Seventeen percent was still not great for the working partners. Jack couldn't have been happier when I told him, but Tom was insulted and backed out of the partnership. That actually worked out better because now there was Jack, my father, Roger and I involved in the second section, and everyone owned 25%.

We connected the new road in the second section to the one running through the first section, then made our first building deal with Mego Toys. Mego licensed super hero figures for boys, such as Batman and G.I. Joe. They wanted a 100,000-foot warehouse. Nervous about committing to such a large building, we went ahead with the deal because they seemed to be a very strong company. Shortly after they moved into their new warehouse, they wanted to build an addition larger than the first at 160,000 feet. I wouldn't attach it because I already had one mortgage and would have to get another. I explained to them that we could build an additional building and had figured out a way to make a tunnel between the two, but I really didn't want to have only one tenant for 260,000 feet. It was too big a risk, given the size of our company then.

I had an idea, and called Steve Curto, a friend I'd met at Pierce, Mayer & Grier. I told him about the deal with Mego Toys for a second building and that I was looking for a partner that would buy half of that building. He introduced me to Ed Blumenfeld, a tall, athletic mortgage broker who had just left Pierce, Mayer & Grier. I had been a mortgage broker there in my earlier days, and we hit it off right away. Eddie bought that deal, and we built a tilt-up building for Mego. Tilt-up construction, a unique concept on Long Island at the time is where you pour cement while on the ground and then you tilt it up into place with a crane so it is vertical. It is an economical method used in the construction of high buildings that saves a lot of steel and time. The success of this venture led to a long friendship and a strong association between Eddie and me for future real estate partnership ventures. Eddie has always been likable, quick-witted, lots of fun, and self-made. He has an innate sense of how far he can push a deal. Eddie could sell you anything and usually did.

Although we preferred to lease buildings and rarely sold them because leasing built up our income, I occasionally sold to friends. As we continued to move fast on the second section, another friend, Richard Karyo, bought a building for his firm. We built Richard a new building and sold it to his company, Sportsatron.

Halfway through the second section, I decided it was important to start acquiring land for the future because I realized it took a long time to file maps and get the land ready. My father had bought land from Maurice Gruber, the number one land dealer in the area who bought directly from the farmers, like a land wholesaler. His father before him had driven around Flushing on a horse and buggy, visiting farmers and eventually buying their land because they trusted him.

The method sounded like a good idea to me. By going directly to the farmers I would be able to get the land at less money. Not many developers had done that. Land on Long Island was being bought and traded and never built. Like the old farmers' story of the bag of potatoes: people kept selling it and making money and selling it and making money. The only one who got in trouble was the last buyer who opened up the

too costly bag of potatoes and found they were rotten. My
father felt the same thing could happen to land and we had to
be sure that if we bought a piece of land, we could build on it.
We were not in business to be land traders.

I started dealing directly with farmers, not an easy task,
since there were only certain times I could see them. During
harvest season, farmers started work at the crack of dawn and
didn't come in until dusk. There were five or six farmers that I
felt were critical to the area, and once every two or three weeks
I made a trip out to see them. I would go into their barns, sit on
a bale of hay or a bag of potatoes and talk with them.

An elderly widow—about 90 by the time I bought the
land—owned the Klein farm and was renting out the land to
farmers when I first met her. I would start my rounds with her
mid-morning, visit in her kitchen and have a cup of coffee. The
farmers wanted to know what the prices were, what was going
on and what the whole idea was before you could even offer to
buy land. They asked a lot of questions and I had to earn their
trust.

I started out by telling them I was a builder, and I
thought their land was the best piece in the area. This wasn't a
new game to them because these were the same farmers who
had sold farms in Flushing, bought in Hicksville or Jericho and
farmed there, then when the market was high, they sold that
and bought a piece of land in Melville. They would break even
farming, but when they sold their land they would make real
money.

I learned that even if I had to say something I thought
might hurt their feelings or our chances, such as what the price
really was or wasn't, it had to be the truth. One thing farmers
could always do was tell when people were lying. They grew to

feel comfortable with me, but I had to wait for whenever they were ready to sell their land.

The town of Islip, the politicians and the planning board loved what we had done with AIP. Adjacent to the rear of AIP 2, there was another large piece of land, about 120 acres going all the way through to Sunrise Highway. Just like our first piece of land, there were homes on the other side of the street but nothing on our side. It was a clean piece, flat, no problems. The town thought it was a good idea to develop it and encouraged us to go in for the zoning. This was our first experience with zoning, but since the supervisor, the planner and the town board were all for it, and it was the continuation of the prize-winning AIP, we *thought* it was a sure thing.

When we arrived at the Town Hall for the hearing, we faced what appeared to be an hysterical mob ready to lynch us. People were carrying on that the development would bring in people who would steal from them and rape their daughters. The lawyer kept telling us, "Don't worry. Stay calm." The whole time the politicians didn't open their mouths or do anything to control the mob. We were shocked and disappointed. These were people we supported, people we thought were our friends, and they just let this go on without saying a word.

"*What we least expect generally happens.*"
— *Benjamin Disraeli.*

The last straw for me was when a blind lady started banging her cane on the desk at the front of the room and yelling, "All their big trucks are going to run me over." I leaned over to Roger and said, "Let's get out of here. This is hopeless." We left before the hearing was over because the outcome was evident.

I learned yet another important lesson at that hearing. No matter how friendly politicians are, if they are in the minority and supporting you will cost them votes or prestige, they aren't going to back you, no matter what they say. Years later, that same lesson would be repeated in our dealings with EAB bankers regarding the Omni building.

We left the hearing feeling very angry. We had done a lot of work, spent a lot of money, brought a huge tax revenue—and good publicity—to the town and had just been called all sorts of ugly names for our trouble. In the car driving back to the office, I said to Roger that it really didn't make a lot of sense for us to build in Islip anymore. If we were going to have to go through this kind of effort we needed a better risk/reward ratio. If we had to put up with the headaches, then the payoff had to be worth it. I suggested we should start looking into Huntington where I had been talking to many of the farmers.

My father and Tom Manno owned a piece of land in Melville, which was part of Huntington, on Bethpage Spagnoli Drive. Soon after the Islip hearing, they sold this land to Roger and me at $36,000 an acre. We had to get the zoning and the variances and the file map on it. While that was going on, we were still in the process of finishing up the second section at Airport.

Our next zoning hearing was in Huntington, and we could only hope that would be more successful. Melville had allotted single tenant use for every building instead of allowing multi-tenants and we needed to change that. We had these large lots where it would have been impossible to make a bunch of small streets, so we planned instead to build large buildings and divide them up for tenants. The variances didn't allow this because Melville wanted to prevent what they called small garage-type manufacturers. Nobody would divide up a large 100,000-foot building into such small tenants, but regardless, it fell under the same zoning rules.

In order to build in Melville, we had to go before the planning board in Huntington, which had a reputation of being very tough. After a long time of getting the runaround I finally got a private hearing—thankfully, not a public hearing. I was allowed to bring one person with me to the hearing. I brought Eddie Blumenfeld, who had become a very good friend as well as a partner in some of the things we did. He was well versed in real estate.

I suspected there were shenanigans going on, but I couldn't figure out what, so we decided that Eddie would sit in on the hearing with a tape recorder. The hearing was proceeding fine with some back and forth give and take. Suddenly, one of the members of the board, a public relations guy who'd lived in Huntington all his life, said to me, "We don't want your kind in this town."

I bit my tongue to keep my temper in check (at times I could flare off pretty fast), then replied, "Well, first of all, I already live in the town. My brother lives in the town. We're not outsiders. What do you mean by *your kind*?"

I knew that I had them because I could see the chairwoman, Lucille Mayer, was very uncomfortable. A couple of the others turned red in the face. "What do you mean by *your kind*?" I repeatedly asked, both for the benefit of Eddie's secret tape recorder and to embarrass the man.

"You know," he blustered. "You know what I mean."

The chairwoman interrupted and said, "I suggest we ask Mr. Rechler a couple of questions, and then I think we have enough information."

At that moment, the tape recorder snapped off and started to beep. Only Eddie Blumenfeld, whom we called "Fast Eddie," would be fast enough to get us out of this one. He started to slap his watch and at the same time quietly turned off the tape recorder. After I answered a few polite questions, we left.

Four days later, they gave us the variances. I'm sure the only reason was that they were afraid of an incident because this guy had opened his mouth. All builders on Long Island were either Jewish or Italian, so the members of this particular planning board were anti-growth, especially this public relations guy. We got it passed that day only because they were embarrassed. Fortunately, not all planning boards were this way.

That was one of the first projects we started in Melville: County Line Industrial Center. We were still finishing AIP at the time, about 1974, and we'd just had the first bankruptcy in our park. Our bankrupt tenant had supplied schools with equipment, and they left behind all sorts of merchandise like audio cassette players, large cases for 8 MM films, some projectors, writing pads and various other supplies. We went to the auction, and of course, all the money that was bid went to us anyway because the rent was unpaid and we were the creditors. However, we were looking for a promotion to announce the opening of County Line Industrial Center (CLIC), so we bought everything for six hundred dollars. Seeing the items had given me an idea for a promotional party.

We'd recently hired a very pretty blond as a receptionist for our Bohemia airport office. The most popular program on television at the time was *Mission Impossible,* so we recorded the *Mission Impossible* theme on all these recording machines, along with the message about our grand opening of CLIC and the date of our party there. It was an audio invitation. We even said, "When you listen to this message it will self-destruct."

We sent our pretty receptionist, dressed in a raincoat, fedora, and dark glasses like a secret agent, around to all the different broker's offices. We put the cassette player, along with the invitation tape and a brochure of the park in one of the film boxes, and she delivered the whole package. One funny incident was when she went in and put the package on one broker's desk and when he saw the recorder and the blond he thought his wife was setting him up and started to run away. It was one of my more interesting promotions, and the party was a well-attended success.

One day, not long after we launched CLIC Roger mentioned that he was taking a week off to marry Sharon, our pretty, blond receptionist. He had been divorced for a couple of years by that time, but it came as a shock to me.

I got a worse shock in 1974, as Judy was planning my fortieth birthday party, when my father was diagnosed with pancreatic cancer. We called off the party, and Dad scheduled surgery at North Shore Hospital. The surgeons discovered the cancer was very serious, and though they cleaned out as much as possible, they weren't sure of the next step.

We inquired around and found out that Dr. Warren at Massachusetts General, which was connected to Harvard in Boston, was a specialist in pancreatic cancer surgery. I called him and told him about Dad's condition. By this time in his career, Dr. Warren was lecturing and teaching other doctors and only took on a limited caseload. He wanted to meet my father and see his x-rays.

I went up to Boston with Dad. Dr. Warren examined him and they instantly liked each other. If you had to invent the prototype of a perfect doctor, it would be Dr. Warren because of how he talked to his patients, the way he carried himself, how knowledgeable he was, and what an unbelievable reputation he had. I have yet to meet another like him. Dr. Warren wanted Dad to return to Boston for a special surgery that he would perform. We were hopeful.

"When you are forty, your warranties run out. When you are fifty, your guarantees expire. When you are sixty, you need parts replaced."

— Donaldism.

Maybe I had always been a type "A" personality, or else somewhere along the way I became one. However it happened, I was definitely a workaholic. About the time I turned forty, it seemed as if all my warranties ran out. I didn't feel well and different parts of my body hurt. Strange things were happening to me, both physically and emotionally.

I had a low boiling point and became very quick to anger. There was a particular incident that really made me aware of this. I had met with a large, highly rated firm interested in renting 65,000 feet of office space in the park. They were well on the way to signing the lease, and wanted us to meet with their architect. We had met this guy before and knew what to expect. "Roger, let me handle this meeting," I warned my brother. "Don't get excited at anything the architect does."

The architect sauntered into our office just as I was lighting a cigar. He hadn't even reached the conference table before he announced in a cocky tone: "You guys build like shit." The match burned my fingers instead of lighting my cigar because I was in such shock at his words. The match also ignited my already short fuse. I jumped up, grabbed him around his neck and threw him out.

After I returned to the room and sat down to try to collect myself, my brother said wryly, "I'm glad you didn't want me to handle this."

We eventually completed the lease deal with our new hi-tech tenants, but we refused to work with their S.O.B. architect. The incident was a wake-up call for me though because I was literally and figuratively burnt.

I also noticed a change in my behavior at home, where I was more short-tempered with Judy and the kids. As time went on, stranger things happened. I would fall asleep every day in my office about three o'clock in the afternoon. I would be on the phone and suddenly forget whom I was talking to, or I'd hang up and immediately forget who had called. I felt my memory was failing me at other times too.

I began to seek medical advice, starting with our family doctor, Dr. Henry. After examining me, he said, "Look, I want to tell you it's mental."

"It's not mental. I know myself. I'm not a mental person. It's physical," I insisted.

After three visits, Dr. Henry could only suggest that I see a psychiatrist. I held firmly to my belief that the problem was physical.

My stress level escalated and my symptoms worsened over the next several months. There was no doubt that the stress of my father having pancreatic cancer and going through various surgeries during this time contributed to my growing frustration and stress.

I took Dad to Boston to have his surgery with Dr. Warren and I stayed in the hospital with him. As he was recovering one of us—myself or Roger—was there with him all the time. On weekends we would both be there. During the week, even if we went home or to a hotel at night when he was doing okay, one of us would be back the next morning. MacArthur Airport, directly across from our office, was running flights to Boston then, so it wasn't like we had to drive to Kennedy or LaGuardia.

The Boston hospital was in a very bad neighborhood where they had even had riots, so I always took a cab. One day, the driver asked me why I was going to the hospital, so I told him. Then he asked a couple of peculiar questions: what business was my father in and what business was I in? When I left the hospital that night about 1:30 A.M. to go back to my hotel, there was the cab—the only cab—outside waiting and it was the same driver. I was a little hesitant to get in, but I did. He started again with the questions, but this time I was careful and didn't really answer him and tried to be vague.

Exhausted, I collapsed into bed at the hotel. I was expecting Roger in the morning and then I would fly back to the office; we kept rotating like that. I didn't know what time it was when I heard my hotel room door open. I had a chain on it, so it didn't open all the way, only enough to see the light from the hallway. "Roger?" I murmured as I got out of bed. Walking to the door to release the chain, I saw a hand reach inside, trying to release the chain. It wasn't Roger's hand! I slammed the door on the hand, and it disappeared on the outside. No scream. Then I heard footsteps retreating quickly down the hall. Obviously, it was either the cab driver, or he'd sent someone. Maybe he'd decided we had money or something worth stealing. I never saw that cab driver again.

My father's surgery was as successful as could be, but the prognosis was that it was still a matter of time for him. He was eventually going to die of cancer. When he came home, Dad confided in us things we never knew. We thought he had a very happy second marriage, but Joan was a troubled woman and had tried to commit suicide. He'd put up a good front and we'd never known.

Dad didn't even have a will written. Wanting to prepare his estate, he asked me to find a good attorney. This was 1975, and a number of things happened then.

The economy had started to tank and values were going down. It would work to our benefit later, but we didn't look at it positively then. At around this time, Franklin National went bankrupt. This was the bank on Long Island that my father had used for his entire real estate career. We had adopted it, and in fact, used the same banker, Connie Stevens. I had a very close relationship with all the bankers at Franklin National.

When the FDIC took them over, I had two of the County Line Industrial Center buildings under construction with construction loans from Franklin. The FDIC wanted them paid off immediately. As we were in the middle of building, there was no way we could pay off the loans. All our payments were up to date, the interest was being paid, and there was nothing wrong, but once a bank declares bankruptcy, the FDIC apparently had the right to call in its loans. We had no money to pay it off, so instead I argued with them and the situation dragged on. Facing failure, I realized how important a role luck plays in success. It was just one more thing weighing heavily on me.

I can see now how the stress and traumatic situations that accumulated added fuel to my health problems. There were times then that I would go home at night, and when I tried to get out of an easy chair, I couldn't move my muscles to get up. It wasn't like weakness, but rather like I'd completely lost my motor capabilities. Fearing some serious illness like multiple sclerosis, I went to another doctor. No one found anything wrong.

One day in October of 1975, I was driving home from the office, listening to the World Series on the car radio. After the game was over, I listened to an interview of Dr. H.L. Newbold. He was describing the same symptoms I was having, claiming they were due to allergies.

It took a long time to get an appointment with Dr. Newbold, but in between what was going on with my father and the FDIC, I called and managed to get one. His office was on Maiden Lane in Greenwich Village. He was quite a character—a tall man with a beard, who wore sandals and always had a dog at his feet. It was hard to say what kind of dog because every month or two, when I went back, he had a different dog. My first impression of the doctor was that he was strange. Dr. Newbold had written three books and had been on television, but when I'd ask other doctors about him, a lot of them thought he was a quack. A very interesting guy, he had started out as a gynecologist, then became a psychiatrist, and now just called himself a doctor.

At the beginning, all he did was question me. "When do you not feel well? What are you doing? Where are you then? Is that where you are not feeling well everyday? Are you at your desk when you lose your memory? Do you fall asleep only in your office? What other things bother you? Do you have a drip when you first wake up or do you have a drip after you wash up?"

I'd had a postnasal drip all my life. When I woke up in the morning I always had a drip and in a couple of hours it went away. Dr. Newbold tested me for all sorts of food and environmental allergies. I started seeing him regularly. It took many months for all the testing as he only tested for two of three different things each time. I saw movie actors coming into his office and heads of state. They even flew in a foreign minister of Russia to see him and we were in a cold war at that time. While I was going through the testing process, he would talk to me like a psychiatrist for half an hour each time.

Finally, when all the tests were done, he said, "Okay, we're ready to get you cured." The first thing he did was tell me to get up the next morning and not brush my teeth. I found out my morning drip was caused by a toothpaste allergy. He changed me to bicarbonate of soda and the drip went away.

"The headaches you have every night when you go to sleep are not tension headaches," Dr. Newbold stated. "You read the newspaper every night. You are allergic to ink print." I was amazed at this, but decided I wouldn't read my paper that evening to see if I could go to bed without a headache. Sure enough, Dr. Newbold was right and my headache was gone.

"When you lose your memory on the phone and when you fall asleep at three o'clock it is always in your office," the doctor continued. "Your office has a mold."

"Impossible! I built the office myself, and it doesn't have mold. It's a relatively new building."

The doctor shook his head. "You are allergic to mold and that is what is giving you these reactions. Look for it."

Still skeptical, I went back to my office and asked my long-time secretary and personal assistant, Beverly Thompson, if we had mold. She thought for a moment, then said, "You know, last summer when you were away on vacation with the boys, we had a bad leak right by your desk. They cleaned the carpet, but maybe there is mold. And you have that old attaché. I noticed mold when I opened it."

I opened the attaché and found it full of mold. We ripped up the carpet and discovered mold under it. Dr. Newbold had been right. My headaches, memory lapses and falling asleep in the office all disappeared. Dr. Henry had said that I probably didn't like my work. I knew that was out of the question. It turned out that my short temper was an allergy to tar, a substance found in ink print. I was allergic to certain perfumes and they would also set me off. Certain men's cologne and after-shave get me especially angry.

I had been battling my symptoms for about a year and a half by the time I found Dr. Newbold and the testing was completed. During that time, the only way I could even function was by swimming 80 to 120 laps a day. The swimming brought oxygen to my brain, which helped. Sometimes, after coming home at three or four in the afternoon, all I could do was lie on a raft in the pool, unable to move, while at the same time tense and trying to relax.

Dr. Newbold put me on a multi-vitamin regime with lots of vitamin C, which would clean the foreign and environmental elements that bothered me out of my body. At first, it was a very heavy dose, but I was later able to cut it back as I got better. The doctor had also determined that I had a major shortage of folic acid, a B vitamin. In addition to the allergies I mentioned, I also had severe food allergies, particularly to sugar and white flour, which made me feel tired. Those foods weren't good for anybody, but for me they were toxic. They affected my memory and thinking. My son, Glenn, had some of the same problems and, after experimenting with his diet, saw that he too had been adversely affected by sugar and white flour.

I remained under the care of Dr. Newbold. He had me eating meat three times a day, no white flour or sugar, and continuing with the vitamin regime. I stuck to that diet religiously for almost 10 years. I was euphoric on it. I have since modified most of it—not the vitamins—and continue to do well.

Dr. Newbold's prices were high. He charged $85 a visit, basically for just talking to you for half an hour. In the late seventies this was high. A few years later, I went into his office and asked him why he charged so much.

He said, "I only get the same rate as a psychiatrist, which I was. But the difference is that when I was a psychiatrist, I cured maybe 5% of my patients. Now I cure 75% of them. Most mental and physical problems come from allergies." He insisted even blood pressure problems were usually caused by allergies.

Dr. Newbold was up for a Nobel Prize. My mother's icon, Carlton Fredricks, was his biggest fan and had written the forewords in his books. It was an interesting coincidence that I would find the help I needed from someone connected with my mother's medical investigations. I was lucky that I had been listening to the World Series in my car that day.

Two years had passed since my father had been diagnosed with cancer and he had come through a third surgery. He had even experimented with vitamins, nutritional and other unconventional treatments. I got him some apricot extract that was made in Mexico that he was injected with. That seemed to help him for a while. He fought hard to stay alive, and for a time it seemed as if he might beat it as long as he kept fighting. But he was not doing well and suffered an awful lot of pain. I visited him almost every day. A newlywed then, Roger wasn't able to get there as often, and when he did come, his new wife wanted to leave right away. (That marriage would only last a year.)

One day in January of 1976, I went to see my father and he looked at me very seriously. "I want you to listen to me," he said. "It's too hard now. I went through a tunnel and it isn't so bad on the other side."

I said, "Tunnel?"

"Yes . . . I guess I almost died. I went through a tunnel and then came back. I wasn't ready, but now I am ready. I want to ask you two things," he said. "When I'm gone I want you to take care of Joan and see what she needs." He paused for a long moment, seeming to gather his resolve. "And be here tomorrow morning at eight o'clock. I want to say goodbye, and then I'm going to die."

I nodded my head numbly. I could hear the pain wracking his voice, see the exhaustion in his body, but his eyes showed the strong, unwavering nature of his decision. I still pleaded with him, but to no avail.

My father had just made an appointment with me to die.

There was nothing I could do but arrive promptly at 8:00 A.M. the next morning. I sat at his bedside, leaning forward to hear him. Joan was sitting in the far corner of the room, sobbing. Dad gave me his ring, the rare cat's eye ring he'd always worn that he'd bought on a trip to India with my mother. "I'm going now, son. Goodbye." He kissed me, then closed his eyes and died.

Joan let out a shriek and started screaming like a Sicilian woman in a deathbed scene from the movies. That's how my father left. He put everything in order, or at least he felt he did, then kept his appointment to die.

It was January 20, 1976. I arranged for my father's funeral, not an easy thing to do. When he'd buried my mother, he'd arranged to have the funeral procession go past the house before it went to the cemetery. I did the same thing with him. He joined my mother at Beth David cemetery.

My mother's death had been particularly hard because it was sudden and she wasn't sick, but even though my father's death was expected and he went on his own terms, it was still a difficult time.

CHAPTER 14

Bashert

Meanwhile, back at the office, I was still being harassed by the FDIC over the construction loans and we were still building. Instead of sending an attorney, I went in every week in person while they were threatening to foreclose. This went on for several months. Business picked up and we began to sign tenants, so I was eventually able to show them I would soon qualify for permanent financing and be able to pay off the loan then. Through Ken Goldman we got our permanent loan from Prudential Insurance Company, and finally got the FDIC off our backs. If you look for someone else to solve your problems you'll never be free of them.

We were getting more involved in the Melville area of Huntington. I bought a large pumpkin farm on the Expressway at exit 49 directly from one of the farmers I'd been visiting. Again, we wanted to build multi-tenant office buildings, but the property wasn't zoned and the town was still very difficult to deal with on this issue. We applied for the zoning and finally got a hearing. They made it so difficult that I couldn't see how we were going to make ends meet building this piece of property. I wasn't sure whether it was me personally they were having the trouble with or that it was because Roger and I had never built office buildings in the town of Huntington before (the previous hearing was for industrial).

My uncle had built large projects in Huntington in the past and I thought he might have a better chance. I had about a year left on my contract to get the zoning, so I told him I would flip (sell) the contract to him. I asked for $100,000 and he countered with $50,000. We finally settled at $65,000. Uncle Mort wanted papers drawn up, and we both used the same lawyer, Bernie Kaufman, who had also been my father's lawyer. But we had never used Bernie on both sides of the same deal before.

I met the lawyer at my uncle's office, and Uncle Mort had his secretary come into the meeting. We sat down and Mort immediately started dictating the deal to his secretary. She was asking questions. Bernie was asking questions. I was trying to follow the rapid fire process.

After a short time, Uncle Mort said gruffly, "Come on, let's go to lunch. Better still, let's get it typed up and bring it with us because after lunch I'm going to be busy." That's how my uncle was. About ten minutes later it was typed up, and he said, "Sign it."

I said, "I didn't read it."

"Well, you sat there. You heard it. Sign it."

So I signed it, wondering what the hell I signed.

After lunch, I returned to my office and read what I had signed. It seemed to be okay. A few months later, my uncle approached the town for the zoning, but things didn't work out any better for him. It turned out that he assumed he had bought the property *conditional* to getting the zoning. Naturally, I wouldn't have sold it conditionally because if I'd thought we could get the zoning I would have held it for the rest of the year to see. Uncle Mort dictated the deal so fast that he thought he put in a conditional clause. I didn't hear him say it. If he did, both Bernie and the secretary missed it also.

Uncle Mort berated Bernie Kaufman who was the sweetest old-timer you would ever want to meet. Not long after that, Bernie got sick and had to retire.

About two weeks later, my uncle handed me a check for $65,000 and said, "Consider it a Bar Mitzvah present." He thought the mistake was his attorney's, but he paid the money because he was an honorable man—a tough businessman but one who lived up to the deals he made.

In February of 1976, not long after my father died, we took a family vacation to Puerto Vallarta, Mexico with his wife, Joan, her son, Peter, our three boys, and Stanley and Iris Rabinowitz and their children. Joan was sedated during the whole trip; we had to move her in, move her out and move her to dinner. Heavily drugged, she must have been taking tranquilizers and sleeping pills.

The boys had a good time in Mexico. It was a sad time for me, but a good way to release the tension that had been building. We'd taken the boys to Acapulco with my father a couple of years before, so Mexico held a nice memory for us. This time, in Puerto Vallarta we frequented a restaurant called Carlos O'Brien's where they treated the kids especially well. As is sometimes the way with vacations, the most memorable event happened when we were leaving to return home.

Stanley had stayed on to do business in Mexico City; he sold parts for the lighting industry through his family business. I took all the women and children to the airport to fly home. Everybody went through customs, but when they asked me for my tourist card, I couldn't find it. The card was a tissue, but it wasn't in my wallet or my pocket.

"You can't get on the plane," the customs official said.

I was carrying my camera case and I gave Judy the luggage ticket. My whole family was on the plane, but I wasn't allowed to board. The next thing I knew two Mexican policemen put me in handcuffs and took me to a station in Puerto Vallarta. There they took the handcuffs off, but told me I had to wait while they checked with authorities in Mexico City.

I sat on a hard chair in the police station for hours and nothing happened; they'd had no word from Mexico City. The plane had departed at 8:00 AM and now it was almost noon. The sergeant talked to me, but he didn't speak but maybe one or two words of English. I asked him if I could walk across the street, provided I left my camera and camera case there. I finally managed to get my point across because he agreed.

Across the street was a little outdoor art gallery and there I met a Canadian woman who ran it. When she asked me where my car was, I told her what happened. "You can't stay here in Puerto Vallarta," she said with alarm. "They'll put you in a Mexican jail and take away your clothes. You don't want that to happen. You must rent a car and drive to Guadalajara. Go to the American Embassy. You must get out of here."

I had a credit card and a little change in my pocket, but not much else. I bought a piece of art in the gallery using the credit card and left. Considering her words, I quickly returned and asked her if I could use the phone. Stanley, meanwhile, had not left for Mexico City yet and was still at the hotel. He didn't know that I hadn't boarded the plane. When I told him what happened, he asked me where I was and drove to get me right away. By this time it was mid-afternoon.

"This is Mexico," Stanley said. "You've got to pay them off."

"They're going to put me in jail for losing a tissue? I'm not going to offer anyone a bribe." The situation again reminded me of the old Abbott & Costello joke. "Besides, I don't even have my wallet. I've got a credit card but no money. Judy went home with all my money."

"Give them something and see what happens," Stanley advised.

"All right. Just give me a hundred pesos," I said, not wanting the amount to be too big or to look like a bribe. I went into the sergeant and put the money, equivalent to about $8.00, on his desk. "Buy everyone lunch," I told him.

As I turned my back, the phone rang. "Señor, they found your papers. They found your name," the sergeant said with the phone in his hand.

I thanked him. He explained that even though they had found my name, I would still have to spend the night at my hotel under house arrest.

They posted a guard in front of my room to make sure I didn't leave. I ordered room service—five bucks—and in the morning they took me in cuffs to the airport. I had to meet my plane at Guadalajara, the only connection I could get, where I changed planes to fly to New York. They removed the cuffs after I boarded the plane at Puerto Vallarta, but when I landed in Guadalajara another cop was there to put cuffs on me again while I waited for the next plane. Everyone stared at me as I sat in the waiting room wearing cuffs, embarrassed, so I felt I should explain. "I only lost my tissue," I told almost everyone.

They finally took the cuffs off and I was on a plane heading back to the United States, breathing a sigh of relief. I don't know why I reached into my camera case for a tissue, but as I did, I pulled out my tourist card. "Oh shit!" I exclaimed. Those sitting nearby who knew the story said in unison, "He found it!" After that trip I vowed I would never return to Mexico and to this day I still haven't.

Dad was 66 when he died. For his 65th birthday, which was in May a year and a half earlier, Roger and I had given him a 1965 Silver Cloud. Already a vintage car then, it is *the* classic Rolls Royce today. He cherished the car and was very proud of it. After his death, it was sold for the estate.

After Dad died, we saw Joan regularly and Judy kept in close touch with her. One day, about three months after he died, Joan went to Aunt Ruth's (my father's sister) house, wearing a tattered housedress and asking for movie money. Joan was a very attractive woman who looked after her appearance, so this wasn't like her. Also, she had all the money she needed. As I'd promised my father, I distributed regular funds and took care of her. She knew that all she had to do was ask and we would send whatever she needed right over without question. However, because my Dad had told her: "Donny will take care of you and get you anything you need," she had begun to believe there was some sort of hidden treasure.

The day she showed up at Aunt Ruth's house looking like hell, she was hysterical. Instead of calling me to ask what was going on, Ruth allowed herself to get riled up, and called her brother, my uncle Mort. He also didn't call me to ask what this was all about, but rather hired the finest estate attorneys he could find for Joan to come to her defense. This was especially awkward because Peter, Joan's son, was working for us as an architect at the time.

This "investigation" went on for several months and was very difficult for all of us. My father always lived well and had a nice sized estate, but apparently Joan thought it was bigger than what it was and wanted it for herself. On the other hand, my uncle had done extremely well, but he ran a different type of business. In office real estate, his business was on a bigger scale, while my father had been in industrial real estate. Being naturally cautious, my father had always had more partners involved in his deals, so he had more going on but owned a lot less of each deal.

He'd known for two years that he was dying, so he'd set up his whole estate with a very good attorney. He'd divided his estate equally in thirds: one third for Joan, one third for my family, and one third for Roger's family. I couldn't understand what my aunt and uncle thought Joan was contesting. I felt part of the action stemmed from Mort's curiosity to know how well my father had done. He had always been very competitive, especially with his older brother.

After about eight months, the lawyers concluded that there was no reason to contest the will and concluded that they didn't want to handle the case any more. Mort, who had been at the meeting, then realized there was nothing more to be had and he let go of it. Unhappy with the result, Joan continued by herself—she got another lawyer and another accountant. This went on for almost eighteen months with a revolving door of lawyers and accountants until finally her reputation preceded her and no one would agree to take her case.

My father's estate was a $6,000,000 estate. It was a substantial estate, but for some reason Joan thought there was a lot more. Years later, Uncle Mort apologized. I'm sure it was the first time in his entire life he made an apology and I knew how much it took for him to do it. Aunt Ruth basically never apologized. She just mentioned how Joan had deceived her. Joan's accusations and behavior had put my whole family through a heck of a time.

Even during the period that Joan was contesting the will, I felt an obligation to continue taking care of her because my father had asked me to. Up until October, which was five months after the lawsuit began, we were still seeing Joan, despite the bad feelings. Judy called Joan and invited her for Breakfast on Yom Kippur. Joan said she didn't know if she would be able to make it, that she might be spending the holiday with a friend, but she would call Judy back to let her know. We never heard from her again. Judy, usually the one to initiate, got tired of chasing her and that was the end of any verbal communication. The only times I heard from Joan in the years that followed were when it was through an attorney or an accountant. She died in 1994.

My father had given her a letter, instructing her to pass certain family heirlooms down to Roger and me when she died. One example was my mother's monogrammed sterling that my grandmother had given my mother as a wedding gift. When Joan finally died, I called her estate attorney and told him we would pay above market value if necessary for our family heirlooms. I named some items, like the sterling and some art my parents had bought that I knew my father had wanted us to have. I offered to pay generously above their appraised value and asked the attorney to get back to me with the prices. We never heard back, and they just sold off everything. I was very close to Dad when he wrote the will and he told me about his letter to Joan and his instructions to her of his final wishes. She left no provision to honor that letter and the lawyers either didn't know about it or didn't care. Roger and I would like to have had a couple of things to pass on to our children.

Dad's timing was as perfect in death as it had been in life. The economy was very bad at the time he passed, so his real estate was devalued and that was actually good because it meant our income tax on the inheritance was lower. The other thing that helped was that since he never owned more than 50% of anything, he got a major deduction for minority interest. Given what they might have been, the estate taxes were minimal.

Dad left me two things that I cherish: the cat's eye ring he took off his finger and gave me before he died, and the advice he'd always given me: "Anything worth doing is worth doing with class and integrity."

In 1996, two years after Joan died, an opportunity fell in our path. We had since bought a condominium and then a house in Florida. For the past 15 years we had spent every Thanksgiving in Florida, with or without the boys and their families after they married and moved away. That year, because Glenn and his wife, Tracy, were expecting, we stayed in New York.

I clearly remember it was bitterly cold and pouring rain that Thanksgiving weekend. On Saturday, we went into the city because I was interested in a particular art auction. I mistakenly thought it was at Christie's, but it turned out to be at Sotheby's. However, since we were already near Christie's and it was raining hard, Judy and I ducked in to get out of the rain. They were having an auction of Latin American art, something I had no real interest in.

My father had collected Latin American art, and had bought two of his best pieces in Mexico, directly from Mrs.

Orozco (one of the big three Mexican Revolutionary artists: Siqueiros, Rivera and Orozco). He'd had to smuggle the paintings out of Mexico back then because the works of these three artists were regarded as national treasures and he was afraid the authorities wouldn't let these pieces out of the country. He bought them right after my mother had died. These were two of the items we had asked to buy from Joan's estate.

Judy was sneezing, so I suggested that we sit in Christie's until the rain subsided. We sat down, and I bought a catalog just to see what they had, remembering my father's interest in Latin American art. It had been 15 years since he'd passed now, and I never followed this kind of art. I was more into modernists and contemporary art.

It was day one of the auction and I started absently flipping through the catalog. All of a sudden, the book fell open to a page that almost made my heart stop. There were my father's paintings!

I got chills up and down my whole body. We'd had no idea what we were really doing in New York, or why we'd gone to the wrong auction house. This was no coincidence. There was a reason we were there at that moment, and that was how I ended up buying back both paintings. I gave one to Roger and kept the other myself. The moral of this story is that coincidences happen all the time . . . or is it *Bashert*?[1]

During the 1970s, we started to travel more. A particular trip that really stands out for me was one we took in the early seventies to Puerto Rico with Stanley and Iris and some other friends. We stayed at the Conquistador Hotel, and one day we all had massages.

---

[1] Yiddish: "it was meant to be, or it is destined."

Stanley and I went to the first session, put our clothes in the locker, and took our massages. Norman Rothstein and Bruce Tucker, both longtime friends, were following us the next hour. As we came out, they greeted us with their towels around them, heading in for their massages. Stan and I went to our lockers and to our surprise discovered our clothes were missing. After searching almost every locker in the room, we found our clothes where Norman and Bruce had hidden them in a closed locker.

We weren't going to let them get away with that. Amazingly, they had left their own clothes within easy reach. I discovered a door that led to the women's locker room so I pulled a "Hansel and Gretel" and left them a trail of shoes and socks leading into the ladies' area. The women's locker room was empty and there was a lock on our side. I felt sure I could lure them into the women's locker room, where they would then find the rest of their clothes, but by locking the door from our side force them to walk out in front of the receptionist from the women's side. Stanley and I baited the trap with a trail of shoes and socks, then we put their clothes through the door. We hid ourselves on the men's side.

In walked Norman and Bruce after their massages to discover their clothes missing. They started to pick up the shoes and socks . . . and through the door they went, right into the women's locker room. As they walked through the door we locked it behind them. Bruce was wearing nothing but a towel and Norman was stark naked, both carried their shoes and socks. We were about to run out to the lobby to watch for them when all of a sudden we heard screaming. The women had come in from a swimming exercise class. We rolled on the floor with laughter.

Judy and I liked to travel, but we would only go away for a week at a time because we really had no one to stay with the boys. We originally had a nurse look after them when they were little, then we hired a professional sitter, Mrs. DeWall. This went on for about three or four years and we were comfortable with her until the boys reported to us that she walked around drinking tomato juice all day long. Tomato juice? That ended Mrs. DeWall's stay with us.

During the seventies, we became very friendly with the Yaffes. We took several trips to France with Bob and Bev Yaffe, and also went to Austria and Hungary with them. The only thing I really liked about Austria and Hungary was the *sacher tortes,* chocolate cakes with a very thin raspberry filling. I really disliked the people. I felt hostility carried over from the war years.

In the mid-seventies, the four of us got a week long tour of Athens, Greece at a special rate because we went on a charter flight. However, we didn't like the second-class hotel the group was staying in. We stayed in Athens for the first two nights, then made our own arrangements and went off on our own to a beautiful new resort about an hour away called the Astir Palace.

Before we left Athens, Judy and I visited a museum. We came out to find a well-dressed group of students lining the large set of beautiful marble steps (like in front of the Metropolitan Museum in New York). As we walked through the crowd, down the steps, I remarked to Judy, "Why aren't the students like this in our country? Look how nice they dress here and how they respect school."

We got about a block and a half away when we heard screaming and yelling behind us. Turning around, we heard,

"Down with Americans. Kill Americans," and saw those same students racing through the streets. I grabbed Judy's hand, and we ran to the corner of the main street. Luckily, there we were able to flag down a cab. As we jumped in, the students, who must have overheard us and knew we were Americans, rushed the cab and tried to overturn it. Somehow the driver was able to beat them off and drove us safely through the crowd.

We left Athens to spend the next five days at the beautiful resort we had chosen. When our week was up, we returned to Athens to meet the tour group at the airport to fly back home. As we drove into the city, the streets were empty. Nobody was in sight. The restaurants were closed. Even Hertz was closed when we went to return our rental car. There was no place to return the car because everything was closed, so we parked the Mercedes at the curb in front of the airport.

Inside, the airport was jammed and all the ticket counters were closed. We finally found our group, and to our amazement, they told us about the revolution in Athens—their hotel had been fired on and they'd slept for four nights in the hallways on the floor and had hardly eaten any food. Meanwhile, the four of us had been wandering around the country having a great time, totally oblivious.

No one knew how we were getting a plane home. The revolution was anti-American. Bob and I looked out and thought we saw our plane on the tarmac. "Let's walk out there and see if the pilot is there," I suggested, so Bob and I walked about 100 feet when, all of a sudden, shots were fired over our heads. It was more likely the authorities protecting the planes from sabotage than revolutionaries, I figured, as we turned around and headed back inside. Just then, a Jeep full of soldiers sped toward us to make sure of our direction and intention.

We sat in the airport all day without food. All the vending machines were now empty, and between the four or us, we had only had two and a half candy bars stashed away. Finally, at two in the morning we were loaded onto a bus and a police escort—no lights or sirens—accompanied us to a hotel about 20 minutes away where we spent the rest of the night. Very early in the morning they picked us up, brought us back to the airport and flew us out. As an unwitting tourist, I had again been caught in the middle of a revolution.

Later, we took another trip with the Yaffes, this time to the Scandinavian countries. This was also a charter flight, but the hotels were more upscale, so we stayed at most of them with the group. Judy and I collected things on our travels. On this trip, we collected glass in Sweden and ceramics in Denmark. We often bought bulldogs wherever we went because we had bulldogs at home and loved them.

Bob and I always had a good time traveling together. We enjoyed experiencing different cultural traditions. In Sweden, we tried the sauna because that is a big thing in Scandinavian countries, but we never really learned to appreciate it. At the hotel sauna, some Swedes pointed out a pool and motioned to us that we should go for a swim before going into the sauna. Okay, when in Sweden . . . .

I dove in first. As I hit the water I thought my heart was going to stop. It had to be forty degrees, like diving into ice water. Somehow, I managed not to give it away because I didn't want Bob to miss out on this cultural adventure. I let him dive in none the wiser. He let out a mighty shriek. Laughing, we headed to the sauna.

To our amazement there were men *and* women inside, not wearing any bathing suits. They wouldn't let us in until we took our suits off. Again, when in Sweden . . . .

I noticed a couple of pretty blond Swedish girls sitting on the cedar benches, but there were also big German men, old, young and even scrawny people in the sauna. Finally, everyone else funneled out and Bob and I were left in the sauna with two pretty girls who were speaking Swedish to each other.

As a joke, Bob said to me, "Reck, look at these girls. Look how they're built. Want to jump them?" I turned red, not from the heat, but from embarrassment. I'd been all over Sweden in the shops and had yet to find one Swedish girl who didn't speak perfect English. I tried to change the subject, but Bob ignored me and kept going on and on.

When the girls got up to leave, they smiled and in the very best Queen's English said, "Have a very good day, gentlemen." Bob turned crimson to the roots of his hair.

During the Christmas holiday in 1977, our family and the Rabinowitz family went to Club Med in Guadeloupe. It was lots of fun and the boys loved it. With some other families, we held a kind of Olympics, playing water polo, volleyball, and other games against the French and Canadians. We had our own team of seven—the U.S.A.—because we had five boys, plus Stanley and I.

Club Med in Guadeloupe is French and French women rarely wear tops on the beach. That trip, Glenn took windsurfing lessons from a buxom French instructor. Glenn was about 15 at the time, and although he took his lessons from the French beauty every day, I don't think he heard a word she was saying.

CHAPTER 15

Innovation

and Business Strategy

Our company was growing substantially by 1977. After our bad experience at the hearing in Islip, we had decided that building office space further west would be a better idea. The Reckson offices were still located in the east, in Bohemia at AIP, but we now began to spread west, building at an increasingly fast pace. That same year, my brother-in-law, Norm Berlin, was hit hard by the downturn in the market. He wasn't anxious to go back to Wall Street. As Roger and I were becoming busier, we needed help leasing the buildings. I asked Norm to join us and to start handling the leasing of our industrial portfolio.

About 1978, my friend Eddie Blumenfeld, who had secretly recorded our Huntington hearing and who had been a partner in the Mego Building, started working with us again. We knew of a large "white elephant" in Jericho, owned by Grumman. Grumman had engineered and built the LEM, the first vehicle that landed on the moon, in this building. After the project for NASA was successfully completed, Grumman had abandoned the building and it sat empty for six years. Still set up for engineering, the building was 100,000 feet and their asking price was now at $3 million. It later went down to $2.5 million.

At that time, for a piece of office land on Long Island to become economically viable you needed to buy it between $15 and $20 per square foot of building. The Jericho building at $25 per square foot was way overpriced and consequently had been on the market for a long time. We thought it would only be good for what Grumman used it for, an engineering building for research & development.

With his usual bulldog persistence, Eddie kept negotiating and started to bring the price down further. One day, we met for lunch at a diner, and penciled the numbers on a napkin, figuring out that we could *almost* make it work for the 100,000-foot building. The price was coming closer to making economic sense for us. "You know what?" I said excitedly to Eddie. "Let Roger and I take one more serious look at the building because it's been abandoned for six years. The deal is very tight, but it might be do-able. I want to see if we can use the existing air conditioning and see what else still works."

We went back and revisited the building. It had a 10-foot hung ceiling. We found a ladder, and Roger climbed up to look at the duct work. As he pushed the ceiling tile up he was amazed to find another 15-20 feet above his head. This was a *huge* building and the extra height wasn't shown on the filed plan. From the top of the ladder, Roger grinned at me as we both realized we could make it into a two-story building and double the footage. "Quick, close the tile and come down," I said. "Let's tell Eddie to just stop negotiating and buy it."

Eddie closed the deal and we bought the building—Roger and I owned 50% and Eddie owned the other 50%—for $1,377,055.50. I'm relaying the exact price because only Eddie could negotiate a number like that. Not only were we able to make this into a two-story building but we were able to get a five-acre lot separated from the property and put another 100,000-foot building in there. However, we would have to go through a variance in order to build the extra 100,000-foot building.

Feeling we had a fifty-fifty chance of getting the variance we needed, I went with a lawyer and made a presentation to the Oyster Bay Town Board. As I spoke I noticed the members of the board were passing a paper around and smiling at me. Seeing this made me feel very good about my presentation. As I walked back to sit down, my lawyer leaned over to whisper something to me and I smiled, thinking he was about to tell me what a great job I had done. Instead, he said, "Zip up your fly."

The variance was approved, perhaps because they felt sorry for me. In the end, we got 300,000 feet of office at a price of four and a half dollars per square foot. It was the grand slam of all real estate deals, a home run right out of the industrial park! We created North Shore Atrium out of this space.

Not long after that feat of luck—or was it the residue of good decisions and hard work, as my father would say . . . probably both—we went to the John Gardiner Tennis Ranch in Scottsdale, Arizona with the Rabinowitzes. While there, I planned the design for North Shore Atrium, a take-off on the atrium office buildings we built in Bohemia. This time I would make one large center atrium with two smaller ones. The large atrium would serve as a pleasant space for people to sit or mill around in, and the two smaller atriums would be gardens.

Having the three would maximize windows and light into the office building.

With new construction the atrium design was perfect for offices, and the design could be used to convert a big, old square building into an office building as we did with the North Shore Atrium. It was the best way to maximize rentable office space, as we had learned with our first such effort in Bohemia.

Part of the North Shore Atrium was a one-story building and there was a large loading dock area in the back which would cost an awful lot to transform into office space. I came up with the idea of using that space for amenities for employees to use such as a gymnasium, and we also put board rooms in there. In our third building on the site we put in a large conference center that seated about 100 people, along with a restaurant.

Our competition thought we were crazy at the time. It took them ten years to copy the idea on Long Island, and in the meantime our amenity packages gave us a huge competitive advantage. When the manager of a potential tenant looked at our building and saw that he could join a health club to play racquetball, or have the convenience of a restaurant right in the building, it was an easy choice. It was a tremendous marketing tool that gave us a real edge over the competition and attracted even more companies to the area.

Putting amenities in office buildings was an idea unique to its time, and we were the first in the country to do it. After the North Shore buildings, we put amenities in almost every office development we built, and still do today.

As the North Shore development was being finished, Eddie and I set up a sales office on site because we had a lot of space to rent. The two of us shared an office and had a lot of laughs during that time. One particularly funny incident happened on a very chilly winter day. After going to Club Med and seeing all the Frenchmen with beards, I had been inspired to grow one, so I had a red bushy beard then. On that snowy morning, I arrived at work wearing a fur coat I used to wear to football games. When I walked into our office, one of the draftsman turned white and fell off his high stool onto the floor. As he turned from white to red, he managed to choke out that he'd thought I was a bear—imagine a bear walking into the office! Susan, Eddie and I cracked up laughing.

The years we spent in the Bohemia location, our staff, Roger and I shared one big office. I was somewhat of a control freak. By sharing an office, I could sit at my desk, or be on the phone doing my own business, and still hear what the seven or eight other people in the office were doing around me. If something interested me, I could get involved, or if I felt a situation was being misunderstood I could correct it. In 1980, we expanded and moved the company to Melville, where we all went into individual offices. I had to learn to let go of some control then. After a while, you get used to it. You realize as you grow that you can't be involved in everything.

It was shortly before this, in 1979, that I went down to Florida to look at property. We had taken some family trips over the years and now with the boys going off to college and having different schedules and vacation times, we felt it would be a better idea to buy a condo in Florida rather than to try to plan a traveling vacation. During my one day in South Florida, between business and family, I was too busy to spend more time looking, but I had read about Boca West and knew a couple who lived there. It was an award-winning planned resort community with golf courses, tennis courts, bike paths, many pools, restaurants, a resort style clubhouse and a health club. Lavishly tropically landscaped, it was perfect for our family vacation home. I bought a Pepper Tree unit in Boca West.

In 1979, we traveled to the Orient with Richie and Arlene Karyo. I was well into collecting artifacts and antiques by now wherever we traveled. I had been forewarned that, while most of the things sold in Bangkok antique shops may be old they are really not valuable, so I shopped with that awareness.

In Bangkok, I bought two six foot bronze Fu dogs, which I thought would look quite unique in front of one of our office buildings. The shop also had thousands of little statues and sculptures—Buddhas and other Oriental figures that they *said* were antiques. I didn't really care and just wanted to make a little display in my house, so I picked out five because they looked pretty. The purchase cost me a total of $1,200. The large Fu dogs, on the other hand, cost about $6,000 with shipping.

When I finally got the artifacts delivered to me in New York, all but the bronze Fu dogs were the correct ones. There were two bronze Fu dogs, but they were less than a foot high. I called the dealer in Bangkok and he apologized, explaining that they must have accidentally switched my order with that of a doctor in Toronto who had purchased the smaller dogs. "He must have gotten yours," said the dealer. "Here is his phone number. Call him up and you can switch them."

I called the doctor in Toronto, but he said he was keeping the larger dogs. When I called the dealer back, he said in his broken English, "Solly. Nothing I can do." I was out $5,500 for the Fu dogs because I figured that what I got was probably worth $500. There wasn't anything I could do, so I chalked it up to experience.

For a number of years, we had those little artifact ornaments displayed at Terrapin, the 14-acre estate we purchased in 1990. In 1998, an art historian and dealer from London, working on the history of a particular artist, wanted to see some of my paintings, so I invited him to come through my collection. While he was doing so, I noticed he kept looking at those little Asian sculptures. I was embarrassed because this man knew real antiquities and must have realized these sculptures were reproductions. "I picked these up in a market in Bangkok," I said lightly, shrugging. "I just liked them."

"Oh, I was wondering," he said. "These two are worthless, of course . . . this one is old . . . not a bad piece, but it doesn't have much value." He carefully examined the last one, then looked up and smiled. "But this little Buddha here is worth about $40,000." Suddenly, it didn't matter that my Fu dogs had been sent to Toronto. This was a real value. It took a while, but I had the last laugh.

On that same first exotic trip to Bangkok, Richie Karyo and I decided we wanted to try Thai massages. When we asked at the hotel desk, they told us the best place was near the American Embassy. The massage house was very exotic and they offered much more than just massages. Whatever you wanted. Having been in Japan I knew they walked on your back, and I thought this would be a great cultural experience. We said no thank you to the rest; we just wanted a traditional Thai body massage. We asked the price, having to tell them again that all we wanted was the body massage. "Okay, but we can't give you a price until you pick the girl." It turned out the masseuses—or whatever the girls were used for—were sold like cuts of beef, prime being the most expensive.

The girls were in a big atrium on each floor and you walked around a hall and looked in to chose the one you wanted. Each ascending floor had higher prices, and as you went up, the girls got more and more beautiful. Thai women as a race are extremely beautiful when they're young. Finally, Richie and I picked number 48 and 56 and we went for our Thai body massages.

First, they bathe you, which we didn't realize until then, so we got a scrub bath. Then they told us to lie down on a mat on our stomachs. We did, and then we found out what a traditional Thai body massage was: they slid up and back in the nude on top of you! So that was our cultural experience. On our way back to the hotel, Richie said, "I don't think I can tell Arlene this, so don't say anything to Judy."

Thinking it was a funny story (not exactly culture shock), I was definitely going to tell Judy, but thought I'd better honor Richie's wishes. That night, we got dressed and met Richie and Arlene for dinner. As I arrived at the table, I could see that Arlene was bright red and fuming. *Oh no, I'm in trouble now; he must have told her* . . . and I hadn't said a word to Judy. I explained and Judy was a good sport about it.

As we were building the North Shore Atrium with Eddie in Syosset, we started negotiations with Mrs. Klein, the farmer's widow I'd been having coffee with in Melville. By then, she was close to 90 years old. We bought her 25-acre property for about $100,000 an acre. Mrs. Klein took back a mortgage with releases, which meant that after downpayment, we (the builder) paid for the land as we used it. The first building we built on the Klein farm property could only be two stories high because we needed variances to make it a multi-story building and that was nearly impossible to get in Melville. This was the first office building we were going to build from scratch—35 Pinelawn Road.

After deciding we would move Reckson Associates offices in there, we wanted to make it special to really feature our talents, but we weren't sure about the design. Even though we had our own architectural department by then, we didn't have a top designer.

Now I never had a problem plagiarizing, but I always made sure I copied the best. It seems that two men I admired felt the same way.

"If there's something to be stolen, I steal it."
                    —Picasso.

"The secret of creativity is knowing how to hide your sources."   *—Albert Einstein.*

There was an office building in Vancouver, Canada that had won all sorts of prizes, which I'd seen featured in an architectural magazine. We did a takeoff on this building, although we used dry-vit as our face material, whereas that building was pre-cast, a much more expensive way to build it. Our new office was completed in 1980.

Because of our innovation with the North Shore Atrium, we had won a national architectural award and were featured in a number of magazines. We were the first in the country to use dry-vit—basically a shaped Styrofoam that is stuccoed over. Dry-vit was an innovative product because it didn't expand or contract with the heat and cold, plus it served as insulation.

Up to this point, much of what we had built had been innovative—the atrium office concept, tilt-up construction, the use of dry-vit, our amenity packages—and we were receiving national acclaim. In the Pinelawn building we tried something no one had done in the suburbs before: to make a building that was accessible around the clock.

Standard practice in the suburban office parks was to install the services so that after eight or nine hours and on weekdays the electricity, air conditioning and heat in the entire building would turn off—or down—automatically and remain that way on weekends. Extra hours were not factored into the rent, nor was there any way to measure them for an individual tenant. We worked with G.E. to create the first computerized electric and HVA system allowing full 24-hour service in a building. It broadened our scope of possible tenants. For instance, accountants worked 12-16 hours a day during the busy tax season, or other professionals such as lawyers might want overtime access to prepare a case. In our new office building, a computer would figure the excess energy used by a particular tenant, then we would bill it back to them. Until that time, not many accountants and similar professionals had come out to the Island or the suburbs. Now they did, and it was another first for us.

In 1980, we moved Reckson Associates into our new offices in Melville and at the same time became increasingly active in the communities where we built. We got involved with area charities and gave back to the communities.

Roger and I opened the new Reckson office with a fully-developed real estate business strategy. We took a little from my father's strategy, and since he had been partners with my uncle, a little of his also. Uncle Mort's philosophy was based on economy of scale and speed of production, which he got from being an engineer in manufacturing. My father was most influenced by the Depression and wouldn't venture far from his own product or his home base. There were advantages to both strategies. My father's was more conservative and safer with less downside risk. Mort purchased better because of scale and high risk reward. More upside.

We developed our own 12-point plan to guide our business strategy:

Limit downside risk. The most important thing in any real estate venture was to limit downside risk, which meant never, *but never*, sign a personal guarantee on a permanent mortgage. (This I got from my father). You would have to sign for construction loans, but never do it on a permanent mortgage. That meant no construction overexposure, no overbuilding. In addition, don't pay too high a land cost and always build for multi-tenant buildings rather than take on a single large tenant that could empty out a whole building and get you in trouble. And last, watch out for bad credit, and never overmortgage a building.

Full service. We became a full service real estate company, so we could control design, construction, sales and finance, which helped with speed and price (as per Uncle Mort). If we provided everything ourselves we could control the price and the quality and get things done faster than if we had to wait for someone to fit us into their schedule.

Reputation. There would always be another tenant, another building, but we realized early on that it was very important to build and protect our reputation. I always told my staff, "Do what's right." We staked our reputation on doing what was right.

Location, location, location . . .   what every real estate person says *is* very important. You always want to find a location where there is a demand, one with good marketability and access to good labor pools. Long term investment.

Quality counts. Pay a little more for better materials or location or amenities because we would be able to get a better price for a higher quality product, and a better return on a longer lasting product. We were building buildings to hold them and lease them. We had to build the best we could, using the best materials, so the building would last longer, because we would own it.

Invest in human assets. When it came to hiring people, whether they be office employees, contractors, attorneys or engineers, hire only the best, even if they cost a little more. It was also important to invest in technological assets, using the latest technology.

Build towards a reserve. We learned this the hard way with our County Line Industrial Park when the FDIC moved in. It would have been good to have a backup plan. At that point, we started to reserve money and we invested with a Wall Street firm, Bernard Madoff, and built a sizable reserve to hold for a rainy day since real estate was a cyclical business.

Build in parks. Cluster your buildings in one area because (a) you would be able to sell more, (b) it would give you the flexibility to move a tenant if they wanted a larger or smaller space and still stay in the same area, (c) managing and maintaining the buildings was easier because you didn't need as many people or to travel as much.

Building flexibility. When we built industrial buildings, we would have more land than we needed and were always looking ahead. If one tenant went out, and the next needed more office space and more parking, we would be able to accommodate the new tenant. We would build a building with extra high ceilings, even if we didn't need it right then because the next tenant might need some warehousing. We put more in the building which allowed us more flexibility for the future. This became important because, as technology came in, we were able to convert more and more industrial space into office or R & D space. Because that was finished space they needed more people and that brought more cars which necessitated more parking.

Superior service. When a tenant calls, do the best you can and react as fast as you can to give them superior service always.

Value added property. We discovered this when we did North Shore Atrium, our grand slam building. If you find a better and more improved use for a building, you get what we called a value added property because you changed the use and increased the value of that real estate, such as changing an industrial warehouse into a multiple tenant office building. Or instead of having 100,000 feet, we turned an existing building with a high ceiling into two stories and doubled our space—value added.

Timing. Timing could trump "location, location, location" in real estate.

CHAPTER 16

Educating and Expanding

By now, Mitch was about to start college. I made only one demand about which colleges the boys could or couldn't go to—none of my sons could go to the West Coast. I made that stipulation because I felt Los Angeles was a wasteland for good people. My feelings about California came about because of something I witnessed earlier.

I had known a banker and another builder who had been very successful, bright young men. When both of them went out to California they got "flaky." I saw each of them a couple of years later, separate from each other, and they'd really lost their edge. Young, excited, aggressive, they'd both left New York for opportunities in California and never returned. They became very laid back and said things like, "Oh, I don't know if I will go to work today. I think I'll play golf." For both of them, this was so completely opposite of their personalities in New York that it scared me.

Mitchell's guidance counselor suggested Emory in Atlanta, Georgia. We didn't know much about Emory, but Judy and I visited the campus with Mitchell and were very impressed. It was a smaller school, but they offered a very good education. He ultimately graduated with a Bachelor of Business Administration degree. Like I had in Miami, Mitch made lifelong friendships through a fraternity while he was at college.

When Glenn graduated high school, he went to Ithaca College in New York, along with Dave Yaffe. He joined us at the condo in Florida for his first Christmas break, but he walked in loaded down with books. At first, I was very impressed. I panicked when I glanced at the titles and saw *Mao: The Rise of Communism*, and *Marxism in Its Pure Form*. "Oh, my God. I'm raising a Bolshevik!" I exclaimed aloud.

Bob Yaffe had been to Ithaca and told me what a beautiful, brand new campus it was. Upon further investigation, I found out that a lot of the professors now at Ithaca had been sixties dropouts during the Vietnam War, and they were anti-everything. I trusted that Glenn was bright enough that he would eventually realize the ideologies he was curious about didn't work. Early on in his life, Glenn didn't want anything to do with capitalism—except the money part of it.

Judy and I always encouraged the boys to bring their friends home on breaks, and to have parties at the house. That way we figured we could see what was going on. The first summer Glenn came home from college, he invited 30 of his friends from Ithaca to a party at the house. I went into culture shock. They weren't like his friends from high school. A number of them were wasted before they even got there and it was a pool party that started early, at three in the afternoon. I didn't know if they had been drinking or were on drugs, probably both, but I made it my business to talk to a number of them.

Glenn was too bright to be in this environment and I told him so the next day, suggesting he transfer to another school. He was very upset. He was passing, but he hadn't been getting good grades in his freshman year at Ithaca. Judy interceded for him, and said he should go another year, raise his grades and then transfer. By his following year, Glenn had picked his grades up. He got accepted to Michigan.

He had met Tracey Weaver at Ithaca, a bright, pretty redhead from Ohio. Although they were just friends in his first year, by Christmas of his second year, they started to date. Twelve years later she would become his wife. Dave, Tracey and Glenn all transferred to Michigan together for their junior year. Glenn did very well there, both socially and scholastically. Michigan is a large school and a lot of the Dix Hills and Long Island children of our friends were there. All the time I heard how Glenn would be lecturing in the quad, raising money for "Save the Whales" or "Ban the Nukes."

The two older boys had attended Half Hollow Hills High School, as Mark did too, but Mark went into their architectural program. It was one of the only high schools that offered two years of architecture and it gave him a great base to get into a college architecture program. He played piano and had his own band in high school.

Half Hollow was a very large school and Mark's graduating class had over 1,000 students. At graduation, he was voted the most popular student, which I have always felt was due to the female consensus in the voting pool. Between his junior and senior year in high school, he met Jackie Gewirtz when they both attended a summer school program at Cornell. Jackie was a bright, attractive brunette from Long Island's South Shore, who nine years later, would become his wife.

In 1984, Mark decided he wanted to go to architectural school. He sent not only his drawings from high school, but music he had written. He only had average grades and average

SATs, but they were very impressed with his music at Tulane in New Orleans. They felt there was a correlation between music composition and architecture, so he got accepted on that basis. It was a special five-year architectural program. I should have put two and two together and understood that Mark also wanted to go to New Orleans for the music. He had a demanding architectural program while he was there. What I didn't know was that he also had his own band for the five years he was at Tulane and that he took piano lessons from some of the top jazz musicians in New Orleans.

In his senior year of high school Mark had too many girls chasing him and was living the party life. After he'd met Jackie at Cornell the summer before, he still dated. Jackie attended college in Albany. She talked to him several times on the phone after he got to New Orleans, and he told her he would call her at Christmas and they would get together. When he returned home for Christmas break, I don't know why, but he didn't call Jackie. After the holidays, she returned to school in Albany, and wrote Mark to say she'd better hear from him right away, otherwise "don't bother calling again, and have a nice life." Jackie told me later that it was a very good letter, one she kept a copy of.

It must have been a good letter. Mark called immediately and invited her to go down to New Orleans and be with him during Mardi Gras in February. From that time on they went steady until they married. In her junior year, Jackie transferred to Tulane. Mark stayed in New Orleans to take his fifth year of architecture, and after Jackie graduated, she went on to take her masters in physical therapy at Columbia.

In 1981, Glenn was graduating from college, Mark was a senior in high school about to graduate, and Mitchell had been out of college for about a year. That year, for a joint graduation celebration we took our three boys on a trip to Israel and Egypt. At the end of our week together, the three boys were to fly to the Greek Islands for a vacation on their own while Judy and I were going to the South of France.

In Cairo the boys made us very nervous. It was Ramadan and the first night we got there they asked us if they could go out. During Ramadan, the people basically stay up all night and sleep during the day. The boys went out in the streets and by three or four o'clock in the morning, they still hadn't returned to the hotel. We were going crazy with worry. Finally, they arrived, and it turned out they had gone all over Cairo and had a great time with some Egyptian boys they had met.

About three days later, after all our sightseeing and a trip to the pyramids and Luxor, Arlene Karyo, who had been our travel agent, called to tell us that President Reagan had issued a warning that no Americans should go to Greece. Overnight, they'd had another anti-American uprising and attacked the embassy. The following morning, the boys were scheduled to fly to the Greek Islands. With this news, I wasn't going to let them go.

It was then 3:00 in the morning in Egypt. I woke the boys and we all got dressed. I called the front desk and asked them to get me an English-speaking guide, and we went directly to the airport. I figured there would be a mad rush. A lot of travelers went from Egypt to Greece and vice-versa. I was finally able to reschedule them on a plane to London and Judy and I left for the South of France on schedule. The boys were in London on their own for five days and had a ball.

Judy and I loved the south of France. For eight straight years we traveled through the French Riviera with either the

Karyos, the Yaffes, or Eddie and Susan Blumenfeld. On each trip we would usually visit one other country for the first week and end up in the south of France for our second week. My friend, Richie Karyo, had been born in France and the Karyos eventually bought a condominium in Cannes.

To celebrate our twenty-fifth anniversary in 1982, we traveled with Iris and Stanley Rabinowitz. One of our four stops along the Mediterranean was La Réservé, a very luxurious small hotel that had one of the top restaurants in the South of France. I invited the Karyos and Blumenfelds up from Cannes to join us for our anniversary dinner. Their specialty was Grand Marnier soufflé. Stanley and I had had the soufflé the two previous nights, and it was great. Tonight, there were eight of us, so I ordered Grand Marnier soufflé for eight.

Twenty minutes later, eight waiters arrived at our table, each one carrying overhead a huge soufflé that was good for about eight people. Richie spoke fluent French and tried to explain what had happened. Finally, I ended up keeping two. We had a wonderful anniversary dinner in the company of good friends and a funny story to laugh about later.

By the early eighties, my boys were growing into young men and had started to come into the business. Roger had three sons also (my three nephews, Gregg, Scott, and Todd) so we had six boys—young men!—between the two of us. I proposed to Roger—and he thought it was a good idea—that we make all the boys five percent partners in every deal we did. The six of them added up to 30%. We agreed that as each son came into the business, they would get an additional five percent, thus bringing their share up to 10% as a working partner. It was a good plan, and the name remained fitting. It was always

Reckson Associates as my father had named it for Rechler's sons.

One thing my father taught us was that everyone had to be paid equally, no matter what their duties. This is very counter to our culture—what if one son isn't producing or isn't as educated or isn't as valuable? Roger and I felt that after two or three years in the business, each boy's income should be brought up to a certain equal amount.

Mitchell was the first of our sons to come into the firm in 1981 when he graduated from Emory. He followed in my footsteps in leasing and marketing and dealing with tenants. It took a few years to gain experience and confidence, to get a number of deals and wins under his belt, but as his confidence grew, he was terrific.

By 1981—my father had died five years before and I had been the managing partner for our side of the family for Vanderbelt Industrial Park, his partnership with Walter Gross—I got a call saying that Walter wanted to meet with all the partners. Roger and I together with Joan owned a third of the company, Tom Manno owned a third, and Walter owned a third along with Howard Rose. At the meeting, Walter informed us and gave us the paperwork showing that he had an offer for the entire real estate portfolio. It was 20 million dollars, and he was recommending that we sell.

The economy was down then. "I haven't looked at all the figures," I said, "but I feel that since you would be selling at the lowest point in the market, it is a mistake to sell." They were all getting older by then, closer to my father's age, and they were dead set on selling. I couldn't talk them out of it. As I left the meeting I asked if they would wait and let me take the papers and spend some time going over them to see if the numbers made sense. They agreed and I took the papers with me.

Judy and I were scheduled for a week-long cruise from Portugal to Barcelona. I sat at a desk in my stateroom every day of the cruise, going through the deal, building by building to determine their value and the value of the entire portfolio. In the end, I felt strongly there was more value in that portfolio than they were getting. At 20 million over mortgages (over the cost of mortgages) it was a very good buy for someone.

It was a difficult time in real estate then, and there wasn't a lot of money available to buy even if I'd wanted to, but I spoke to our friend and mortgage broker, Ken Goldman. He thought we could get the Dime Savings Bank, which had supported us at AIP to put up 10 million dollars if Reckson put up three and a half million (because we already owned a third, we only had to buy the remaining two thirds). That way, we would be able to top the other offer and buy the portfolio.

Dime Savings agreed to lend us the money, and I brought the offer to the other VIP partners. They looked at each other, waited a few minutes, then changed their minds about selling. I was happy because we held onto the portfolio which was my father's legacy. The next year, in 1982, the values started to shoot back up. We never sold my father's real estate portfolio, and in fact, we still have it. We eventually bought the other two thirds of VIP and now own all of it.

Meanwhile, back at Reckson's building project in Melville, we ran into an unexpected financial problem with 35 Pinelawn. When we had started that building in 1979 the interest rates were 10% and when we completed it later in 1980, we were into the Jimmy Carter years and the interest rates had risen to 16%. Initially, we had figured our profit based on the lower rate, and now the deal didn't look so good anymore.

Eddie Blumenfeld was a one-third partner in that building, so we talked it over and decided to offer it for sale to the mortgagee, with Reckson remaining as property manager. We sold it for $100 a square foot which gave us the original profit we had penciled at 10%. It was an interesting building, kind of like the overpriced bag of potatoes I mentioned earlier. Ten years later the mortgage company, John Hancock, sold it for $135 a square foot, and then it was sold for $160 a square foot. In 1996, we eventually bought it back for $75 a square foot.

In 1981, Bob Yaffe, Dave Abel and some other partners formed a company called United Realty's Partnership and bought a piece of land across the street from where we were building in Melville. We partnered with them to build a small 110,000-foot building and Bob's firm rented an office in it.

The difference between having people like Bob or Eddie as partners, compared to the partners my father had put us with, was they were *working* partners; they were in real estate. Bob was a broker, and Eddie was now becoming a developer, so they brought more than money to the table. In the partnership with Bob we also built a one-story building, similar to our atrium buildings but larger, on a piece of land we were able to buy on Motor Parkway in Hauppauge. It was a successful building.

After we sold Pinelawn to the insurance company, we bought a few smaller deals, then Reckson started to build 225 Broad Hollow—our largest building to date on the Klein property. It was a 165,000-foot building on Route 110. Eventually we won all sorts of architectural awards for original design. In 1982 when it was completed, we moved our offices in there.

As always, while one project was being completed we were looking around for other opportunities. I bought 400 acres similar to the McKay Radio site my father had bought when they built V.I.P. in central Islip, just because I thought it was a good buy. I bought it from Chase Manhattan at $2,500 an acre with only $25,000 down. The problem was that it was a big piece of land mixed with pine barrens, which meant I would have to get involved with the New York Environmental Department. There were road systems to put in and it had a State Highway on one side. The border between two different towns crossed it, which meant I would have to negotiate and file maps with two planning boards. Overall, it was going to be a lot of work and take up most of our time.

About this time, Bob DeFazio, our longtime friend and electrical contractor, let us know that he'd heard that bids might be taken from builders to lease land at Mitchell Field, which had been a United States Army-Air Force base during World War II. There was dedicated parkland there now, and a very large area where one builder had started to build a high rise like New York City. It was a terrific opportunity for us. Located in Nassau County, far from where we were in Huntington (the most westerly part of Suffolk County), it would give us a totally different market. We could build two projects at the same time without compromising any tenants because there was no crossover; we would be marketing to completely different tenant needs.

We began a major negotiation for Mitchell Field. Jack Kulka and other builders were competing with us. Jack had come on the scene 10 years earlier and I felt he was more of a contractor than a builder—a merchant builder really because he would build to sell. His reputation preceded him, but he had a great sales pitch. A manufacturer or warehouse business would think they wanted to own their own building, and Jack made his package sound appealing, but in the end it always cost much more than he said.

While competing for a tenant at AIP, I heard a rumor that Jack was badmouthing Reckson, but it later would come back to bite him. Jack had tried to pressure a tenant that I had been negotiating with, and that tenant made a tape of his conversation with Jack, which he played back for me. Jack said on the tape that Reckson didn't know how to build, that our buildings leaked and we were always calling him for help because he was a better builder. It was total fabrication.

I was still young then and it was pre-Dr. Newbold. I simply flew off the handle. I had a baseball bat in the back of my car, and I drove to Hauppauge looking for Jack. Luckily, for him and for me, I didn't find him.

Like the famous nemesis that Sherlock Holmes imagined was behind every evil deed, Jack was the first of what I called "Moriarty." Through the years, as we grew more and more successful, we always had our Moriarty, whether it was a competitor like Jack, or a broker, banker, or market analyst. The faces of Moriarty changed, but often for reasons unknown, someone would just come after us. Fortunately, as with Jack, it didn't last long. I quickly realized that the way to deal with Moriarty was to get mad, get over it and move on. *Sour grapes make lousy wine.*

Even though Bob introduced us during the Mitchell Field negotiations to the Nassau County Supervisor, Fran Purcell, and we hit it off, we still ran into a few snags. Fran was trying to increase taxes for Nassau County, so he wanted to lease the Mitchell Field land rather than sell it. We preferred to buy it, but since it was such a good location, we agreed to a 99-year lease.

Just as we were working out the lease deal, we discovered that someone else was creating an obstacle. At first, we had trouble deciphering who was behind the situation. The industrial commissioner, Cacciatore was very political; we called him chicken cacciatore, and word had it he was on the take. We eventually discovered he had been partying with Jack Kulka and working against us politically.

We finally prevailed and signed the lease on the two sites that I wanted bordering the park. One triangular piece was directly across from the Nassau Coliseum where the New York Islanders played hockey. We would build a large office building called Omni there. Our second site was on the other side of the park, where we would build two smaller office buildings. All together we would have approximately one million square feet of office space.

There were already major roadways through the whole area, and we decided to start with the smaller buildings first. At the same time, we were still building in Melville, so we were now opening two fronts. Our organization was growing and we added Kathy Giamo in April of 1981.

While we were doing all this building, the bank was pushing for a closing on that 400 acres I had put the deposit on in central Islip. Feeling that we were spreading ourselves too thin because we were already building Mitchell Field and Melville, I probably made the worst real estate decision of my life. I dropped the central Islip deal and walked away from my $25,000 deposit. Several months later, that same piece—the one I had negotiated for at $2,500 an acre—was sold to Wolkoff for $7,500 an acre and they built a major industrial park there. It's water under the proverbial bridge now, but had I hung onto it, that land would have kept us in the industrial building market; it's been a very successful park.

By the eighties, things never even slowed down in our business. In 1983, we bought another piece of land. Maurice Gruber called and asked if we wanted to buy their prime corner piece—Long Island Expressway and Round Swamp Road in Huntington—because they were ready to sell. A great piece of land, it wasn't zoned and we already knew how impossible it was to get zoning in that town. When I reminded him of that, he said, "I know. That's why I'm calling you. I feel you have the best chance to do it. I'll give you a little better price than I'm asking from everyone else."

We bought the property subject to zoning. At that point, there was nothing zoned from Route 110 to Round Swamp Road, so we weren't sure if we would be able to get the zoning. Route 110 is the major corridor, just after Nassau County ends and Suffolk County begins. Crossing both the Long Island Expressway and Northern and Southern State Highways, Route 110 connects all the east/west routes going into New York City.

Hoping to avoid our previous experience with the lynch mob, we contacted every expert we could and really got all our ducks in a row in preparation for the hearing. We even brought in the labor union that had many of the people living in the Huntington area and filled up most of the seats with trades people wanting jobs in the new construction. We made sure there weren't many seats left for crazies or what we called NIMBYs which means "Not In My Backyard."

There was residential land on the opposite side of the Long Island Expressway, but none on or near our property. The residential people were the only ones that had even a minor objection. Just as I had with the farmers, I met and got to know some of the civic leaders. There was one woman, Sonia, who was influential in the community, and we became friends. I explained our plans to her in detail and showed her everything. At the hearing she spoke in our favor, and we got the zoning.

We got the zoning for a 110,000-foot building and built an office building like we'd built at 200 Broad Hollow Road, but this time we came up with another concept. In 10,000 square feet we created "instant offices" by setting up each space with phones, desks, copying machines, whatever any good office should have. We hired receptionists and secretaries who could work for several offices at one time. A tenant would then have a ready-made office and a mailing address. Except for his rent, he paid for other services only as he needed them.

Long Island was full of entrepreneurs and start-up businesses, so we began to put about 15,000 feet of instant offices in every building we built. We found these offices were great incubators for future tenants because some of these small firms would grow and become successful. One company started out in one of our "incubators," and ended up being a 135,000-foot healthcare company, a major tenant in a building.

Our instant offices served a multitude of purposes for Reckson. They provided us with a temporary sales office in any park or building, and a place to make a phone call or a copy or whatever we needed at any time. They gave us a management presence in every building and park, and the conference rooms gave us a place to hold a meeting wherever we were. The instant offices became a separate business for us.

In 1983, we started building the corner of Pinelawn and Route 110, which was the last piece of the Klein farm (completed in 1984). We had barely laid the foundation and started construction when Grumman said they would take the whole building. One of our points in our 12-point plan stated that you shouldn't rent to only one tenant, so we had serious doubts about renting only to them. It went against what we believed in, and because Grumman did government work and engineering, they could only give us a five-year lease. Despite our trepidation, the deal was attractive because we could bring the price of the building in, with a single tenant, much cheaper than having to finish it for multi-tenants. The construction was only just being started and it would be rented immediately. Although at this time it was still a little difficult getting financing, we now wouldn't need a construction mortgage, but instead could go right into permanent financing. Grumman's good credit would give us added support. We took the leap and did it. As we thought, it was very profitable in the short term, but after five years, they left.

Grumman vacated all the buildings they had in the Route 110 corridor during 1989-1990, leaving 1,200,000 feet empty. They had been Long Island's largest firm, employing 25,000 people and they went down to 8,000 people. As the manufacturers of all the naval airplanes during the war, and then big participants in the space program, they left behind a great technical base of labor on Long Island. Many of these people eventually would become entrepreneurs and start their own businesses, but that didn't help us now.

When Grumman vacated our building at the end of their five year lease, I went to see Travelers, the mortgagee of the building. We wanted to suspend the mortgage payments and put five million dollars worth of work into the building in order to convert it to a multi-tenant building and re-rent it to more tenants. We weren't asking them to forgive the loan. We were only asking them to delay the payments and roll the additional amount on the back end of the mortgage until we got the conversion underway and were able to re-rent it. Then we would start making payments again.

Banks were nervous about real estate because of the recession and they weren't making new loans. Travelers would have none of it. It was the first and only time this ever happened to us, but because we broke one of our own rules by renting the entire space to a single tenant, we ended up giving them the keys. We knew that unless we converted to multi-tenant it wouldn't be rentable, and we weren't willing to put up the money ourselves to do the conversion and continue to pay the mortgage at the same time. We had made money on it and already come out ahead, so we walked away. A year and a half later, the bank was able to sell it and recoup their investment.

Our plan when we began to build at Mitchell Field was to rent to large New York City firms who wanted back office space where they would be able to get more productive labor at less money than in the city. We had started on two buildings in Mitchell Field. One was a 185,000-foot, six-story multi-tenant building at 50 Charles Lindberg. Next door, at 60 Charles Lindberg, was a two-story 180,000 square foot building, a large footprint with a full basement.

The two-story building went up very fast and was ready to be rented in 1984. The six-story was about a year behind it. We realized that even though the two-story wasn't as sexy a building as the six-story building and we would have to rent it for less, the profit would be higher because it was economical to build. We attracted First Chicago. They were looking for a location for their credit card department. Our contact was through their Manhattan office, but we were competing with Elgin, Illinois because they owned buildings there. "Show us why we should pay $16 for this location when we pay 65 cents in Illinois," they challenged.

They owned an old loft building in Elgin, Illinois that they had moved one of their divisions out of. We didn't know if they had to fix it up, but competing the best way we knew how, we took much the same approach as we did with Met Life by showing First Chicago how good the labor pool was here versus the Elgin, Illinois area. We arranged for them to interview people. As they were concerned about access, we promised to run a shuttle bus to the train station for them. They couldn't fathom how employees traveled all over Long Island in cars because they were used to cities like Chicago with good mass transit. In the end the deal was made. We ran the shuttle bus for a couple of months, but no one rode on it.

We had trouble with the construction end of this project. Over the years, we found a persistent issue popping up at different times with either the construction department or the real estate department. Somebody there would be on the take and would want to have their own sources involved. The head of the construction department from First Chicago insisted they wanted to hire their own contractor to build out the inside of their office. We would have none of that. Finally, wanting to get the job underway, we allowed them to do only their air conditioning on the basis that we took no responsibility for it. If it wasn't sufficient to cool the place it was their problem. There was no doubt in our minds that payoff was involved when we saw what a poor job was done with the design and installation of the air conditioning. Three years later we heard that on a different job this same fellow was being prosecuted—when there is smoke there is usually fire.

In 1985, we completed the six-story building, so we were well underway in Mitchell Field. For some time, we had been working on plans for the Omni Building, which was going to be our *pièce de résistance*, even though our father had warned us: "never build a monument."

CHAPTER 17

Next Generation
and Omni

In 1984, for my fiftieth birthday, Judy had a surprise party for me on the Riveranda Boat that went around New York City at night. It was beautiful. The boys entertained.

Mitch played the piano from time to time, and Mark played piano and even sings professionally now. Mark's first love was always music. Glenn played the saxophone at one time, and he also sang. At my party they were all singing. It was a lot of fun.

By 1985, we felt we were outgrowing the condominium at Boca West. We'd been there five years already and loved the community because it was the first planned resort community, not just a golf club, but a real resort community. They were opening up a new section of homes, rather than condos, and I noticed a lot right on the lake that I liked. However, they only wanted to sell lots in the first section at this time and the one I liked was in the second section. I kept bothering them . . . and bothering them. I became such a pest that they figured the only way to get rid of me was to sell it, but to ask me for $50,000 more because I was picking the choice lot. I agreed. By the time they got around to building the second section all the other lots were the same price—or more—and I had already bought the prize! I designed the house in Mahogany Bend myself and we had it built by a builder. We designed it in a Southwest style because we had started to go to Santa Fe around 1985 and 1986 and collected a lot of Native American pottery. It was an easy resort living style that we liked because it was a vacation home.

In 1985, Mitchell met Debbie Ullian at Club Med in the Turks and Caicos Islands. She lived in New York City and they began dating. The first time we met Debbie was at the wedding of our friend Bob DeFazio's son, Kenny. Mitchell was invited, but although he and Debbie sat at our table, they weren't close enough for us to have a conversation with her. We weren't sure what to make of it then, but it was obvious this was serious.

On August 29, 1987, Mitchell and Debbie were wed at the Plaza Hotel in New York City. It was a very beautiful wedding, and a lot of fun. We could tell they were really happy and in love, which was all we could ask for our son. That was the first wedding we went through as parents of the groom, and there were about 250 guests, which was smaller than our next one.

That same year, we bought an apartment in Manhattan. Judy was born in Manhattan, and had always wanted to go back. She enjoyed going into the city for ballet, art and various things. One day, while looking at the galleries and strolling along Madison Avenue, Judy and I saw a new condo building going up. We bought an apartment at 80th and Madison, which we still have today.

The Christmas before the wedding, we took the three boys and Debbie to Africa. Bob and Bev Yaffe came with their boys (their oldest son, Alan, had just married so his wife, Kathy, was with us, along with David and Rick). All together, we were a group of twelve. We went on safaris through Kenya: to the Serengeti, Amboseli, and the Kenya Safari Club. We made three governor's camps, sometimes traveling in the four wheeled vehicles and sometimes in a bus. Whenever we could, we took small private planes, which all of us fit into with a minimum amount of luggage.

All twelve of us, on our way to Africa, stopped in London for a few days and then proceeded to Nairobi. In Nairobi, our family went to a game restaurant. There Judy, and no one else, ate wildebeest. The next day, with great anticipation, we were to go on our first safari. As we left for the game drive in the late afternoon, Judy said she wasn't feeling well. By the time we were driving through a jungle area, we had just spotted a leopard and were trying to get a closer look, Judy said she must get out of the truck. It was obvious that the wildebeest had bested her and it was an emergency. The guide told us the area was full of wild animals. Nevertheless, Judy insisted. I got out with her to stand guard, brandishing the only weapon I was able to pick up from the ground, a little twig slightly bigger than a Popsicle stick! Luckily, I didn't have to fend off any wild animals and we continued our ride.

The next several days, we went on different game rides in the various areas looking for and photographing the "Big Five." We did see four out of the five, but the real bonus pictures or viewings were when you saw them do something extraordinary. The most exciting view, I felt, was coming upon nine rhinos in an open plain under a tree, looking as if they were having a meeting. One of these animals weighed more than the truck and could knock it over by itself. As they saw us approach, they fell into a formation with the cadence of a well-drilled football team, standing back to back, protecting each flank.

We stayed up a couple of nights watching a lioness hunt around a water hole. We saw a number of kills at another time. The lion's pride would cut off their meal ticket and then attack. We saw a large amount of hippos fighting or dancing in the water. Finally, on our sunset drive, we saw two amorous lions. We also visited a Masai village and went into their huts made out of dung. Our next activity was showering!

Once, when we were in a bus with another group of people on our way to governor's camp, it suddenly occurred to me that I had put my underwear in the night table drawer and forgotten to repack it. In the middle of the crowded bus, I let out a scream that I'd left my underwear in Amboseli. The Yaffe boys still tease me about it.

We enjoyed Florida in the winter and our new house in Mahogany Bend, where we had been for a few years when Judy was about to turn fifty on February 14th I decided to throw her a fiftieth birthday party for the entire four day Valentine's Day weekend. We invited seven other couples. Our closest friends came. Debbie and Bob Stillman, Peggy and Bruce, and Iris and Stanley stayed at our home. Arlene and Richie Karyo, Susan and Eddie Blumenfeld, Bev and Bob, and my brother Roger all had their own places in Boca West. I arranged for a full list of activities, including tennis, golf and a golf clinic.

We went to jai alai on a bus that night and polo on Sunday afternoon. We went to the top restaurants, had a catered breakfast and several lunches at our home. It culminated Saturday night on Valentine's Day with a catered party. The weekend was a lot of laughs, but two of the funniest episodes that I remember involved Bob Stillman, our friend from Boston. They were staying in our media room when in the middle of the night, Bob was startled by a strange noise. He rushed out of the room, not knowing what was going on. I had just installed a fax machine, which was a pretty new item at the time.

The second incident happened the next day when everyone came for breakfast. We were sitting around the living room and Bob got up to go to the bathroom. When he returned, he calmly announced, "You have a snake in your bathroom." No one believed him. I went into the bathroom and sure enough, there was a black snake. I later found out it was poisonous. I picked it up with a towel and threw it outside . . . after all, it wasn't invited!

I enjoyed travel and parties, and I was fortunate to be involved in a business I enjoyed just as much. Meanwhile, back at Reckson Associates . . . .

As each boy came into the business as a working partner he took on different business responsibilities, so there was no major competitiveness. In 1985, Roger's oldest son, Gregg, joined us and followed his father in the construction part of the business. Gregg took to construction like a fish to water.

Mark came into the architecture department in 1989, having graduated from Tulane as an architect. He quickly computerized our architectural department. As each boy came in, they learned not only their end of the business, but Reckson's business philosophy and culture that my father had taught us: whatever you do, do it with class and do it with integrity because nothing is more important than your reputation.

In business you are going to find somebody will do something one way and even though it doesn't make any sense, they will repeat the pattern because that is their nature. A story that illustrates this philosophy was an old African fable that I told all the boys. *Wanting to get to the other side of the lake, the scorpion asked the frog for a ride on his back. The frog said, "I can't do that. You'll bite me and kill me." The scorpion said, "I would never do that because if I bite you, I'll drown. I can't swim." So the frog said, "That's right. Hop on." The scorpion got on the frog's back and they started across the lake. Halfway across, the scorpion bit the frog. "Why did you do that?" the frog said. "We're going to drown now." The scorpion said, "I can't help it. That's my nature."*

The reason I say this is that around 1987 I started to feel our cash flow was hurting. I had developed a feel for the business that I learned to trust over the years. By how we were renting, I knew what our cash flow should be, and it kept feeling low to me in the late eighties. Something was wrong.

After graduating from Clarkson University with a degree in Business Administration in 1987, Scott (Roger's second son) was the fourth son to come into the business. I mentioned the low cash flow to Scott when he took over finance because I wondered if someone was stealing. I couldn't imagine what was going on or why we were short of cash when there was no reason to be. Scott started to look into it.

Our controller, Shelley, had been with us for 15 years, and had helped us computerize. As Scott went through the finances, he discovered that ever since we began computerized billing six years earlier, we had not been billing the extra taxes to the tenants that were supposed to be remitted when we paid the County. This amounted to almost a million dollars per year, so we were down nearly six million dollars over the six-year period. No wonder our cash flow was low. When I questioned Shelley, he admitted that he didn't know how to make the computer work and instead of telling someone, he just kept putting it off. Not wanting to disappoint anyone, he let the computer run him and very nearly hurt us permanently.

In the years that passed, many of the tenants had left us and some had gone out of business. Our current tenants knew they owed the money, so they simply paid up, but others were more difficult to collect from. Contractually, we had a right to it, but we did have to take some of the former tenants to court and settle with lawyers. It wasn't an easy time and it took two years to overcome this mistake. We were eventually able to collect about four million dollars, two-thirds of what we'd lost. I never did find out whether the computer was dysfunctional or Shelley. He was embarrassed to let anyone know he didn't have the expertise, even though we had computer people who could have helped him. Typical of some people, rather than admit to one little flaw or problem, they hide it and the cover-up makes it worse as it becomes a major problem.

"Some people rust out before they wear out."
—Donaldism

Reckson's most ambitious project to date—Omni, the "monument" on Mitchell Field—was slated to be a 550,000-foot building. Every Thursday for the past six years Roger and I had met with design architect Tom Mojo to plan Omni. After six years, you can imagine, we had this building designed to a T, both esthetically and technically. Featuring all the latest, most advanced technology, Omni was to sit on the triangle-shaped piece of land in front of the park and across from the coliseum. I came up with the idea to design the building triangular to conform to the land, like a ten story wedding cake.

We ran into one problem. Fran Purcell, the Nassau County supervisor who had signed our original deal for the ninety-nine-year lease, had left and elections were soon to be held. The assistant supervisor, Owen Smith, had worked out all the details with us and had been very friendly. A total gentleman, Owen was from an old Long Island family that had been on Long Island for 200 years.

Now, suddenly we were running into one delay after another with the county. They decided there was a water shortage and Omni would use too much water, or it would bring too much traffic, or it would cause too many environmental problems. You couldn't get enough electricity or there might be sewage problems. As we adjusted to each delay and fended off each problem, they changed the zoning and said we didn't have enough land to build the Omni building.

We had a commitment from Nassau County and a plan which had been filed for a long time. There was no reason for the variance change. The political climate had changed and the Omni building fell out of favor.

With Owen's help, we finally prevailed. He convinced the county to compromise by bringing in additional revenue which he created by making us sign a lease for an additional fifteen acres. We had no use for the extra acres, as we already had more than the legal parking requirement. Today, the land is used for museums and we still pay rent on it.

In 1988, by the time Omni was finally started, we had lost the market. We'd been building right and left, but now the real estate market started to get a little shaky. Remembering the major importance of one of our twelve points—*Timing*—I got nervous. This was our biggest building ever, almost three times the size and four times the cost of anything we ever built before. It was a $100 million building, and that's the size of the mortgage we would need.

I went to European American Bank (EAB) which had taken over Franklin. The banker's name was Dempsey, a Long Island boy from Great Neck, and I knew him well. I showed him the project and he was interested. He was looking to make big loans and our success rate was terrific. The only thing I was afraid of was another of our twelve points—signing for loans personally. I went to lunch with Dempsey, knowing that all building loans had to be signed personally. "Look, we'll sign for ten million personally," I proposed. "Anything above that, we can't take the risk."

He agreed and the papers were drawn up. Even though the timing was shaky and I was still nervous, we had a lot of money put away in reserve and the banker had expressed considerable confidence in us by pushing the loan through. If we didn't move ahead with the project now we were going to lose our zoning and all the years of work and effort we had put in so far.

As the building was going up, we attracted Chubb Insurance Company. After a difficult negotiation, they took almost 90,000 feet of space, about a floor and a half. Chubb said the way I could close the deal with them was if we made the building in granite. I said, "I'll make it in granite if you'll pay an extra dollar a foot in rent." They agreed to the rent and we closed the deal. In negotiations the people you think are unreasonable are sitting across the table thinking the same about you.

All went well with the construction and we completed the building in 1990. Paine Webber came in, wanting the most luxurious offices they could find, and took an entire floor. Omni was written up as the best suburban building ever. Still to this day, people in the real estate industry say they've never seen another building like it.

Omni was the first granite luxury building with fabulous amenities on Long Island. It had a 25,000-foot gym with an Olympic swimming pool, a 150-seat conference center, teaching rooms and board rooms. It gave us a competitive edge because tenants didn't have to build a board room into their own space or they didn't even need kitchens or lunch rooms because we provided their staff with tables in the restaurant areas even if they wanted to brown bag and not buy lunch. We featured these kind of space saving amenities in the lower level where we wouldn't be able to get rent for most of the space anyway.

The best amenity was a major art gallery where we had shows. Starting with Pinelawn Road, Roger and I used to buy art for the buildings. It became more important with every subsequent building we built, and we put in some pretty good art. The collections shown at the Omni gallery were very special.

Tenants were moving in and the building was commanding top of the market rents—the highest rents ever on Long Island. And then the "fit hit the shan." The economy plummeted like a granite stone dropped from the top floor of the Omni building.

In 1990, there was a recession spreading across the country, but on Long Island it hit much harder. Reports later claimed that this recession impacted New York as seriously as the stock market crash in 1929 that had brought on the

Depression. Before now, Long Island had barely been hit by other recessions, but this was Depression-scale. Grumman had now pulled out and left all the empty footage and a carnage of job loss.

When the recession hit in late 1990 and into 1991, we were still in the process of renting Omni. Rents were going down and tenants were becoming scarce. Empty footage was being left all over New York and we were now competing with sublet space and many other empty buildings. We had empty spaces in other buildings ourselves, but we had this brand new, very expensive building that could soon become a granite albatross around our necks.

We had moved our first two tenants into Omni and then we got a third, Episcopal Healthcare, but it was like pulling teeth to fill that building. Finally 55% leased, the rental income still wasn't carrying the building. More than halfway home, we hated to lose now. We gradually found more tenants for the Omni building, but these were tough times and in order to attract new tenants we had to charge less rent and offer large concessions and expensive work letters. The building was not yet operating profitably.

In 1991 and into '92, we renegotiated our mortgage with European American Bank (EAB). The process took a while as it was still a bad time for real estate loans. Finally, I breathed a sigh of relief, thinking we had put a deal in place that we could live with. We got it all on paper and were just about to sign when they suddenly fired all the people we knew, including my friend Dempsey, the fellow we'd made the deal with.

EAB divided their bank up into what was, at that time, commonly called a "bad bank" and a "good bank." They installed a banker—more like a repo-man (repossession) wearing a nice suit—in charge of the bad bank. No one knew much about Drake, the illusive head of the "bad bank." He had two people under him; both seemed decent enough, but they had little latitude. Drake's goal was to liquefy assets, and whatever they considered a bad loan or bad risk they put into the bad bank. Omni, our prized building, was put into the bad bank.

Our mortgage was large, complicated, and involved six different banks. European American Bank controlled fifty percent of it, and the rest was split between the other banks. In order to refinance as per the arrangement we'd already made, we had to talk to EAB again because they were the lead bank. I realized it was a whole new ballgame. Drake's underlings were giving us the runaround and the bank had no intention of living up to the first deal we had negotiated with them.

As this was getting us nowhere, we took a trip to Japan where I tried to have friends line up contacts with Japanese banks who might be interested in our mortgage. When I left for Japan, I was expecting to hear from someone regarding an appointment, but the arrangements fell through after I arrived. The day I flew back from Japan, I finally got a meeting with Drake.

Sick with bronchitis and jet-lagged from the time change, I flew right into Manhattan and met the boys and Roger at the bank. The "boys" were Mitchell, Roger's eldest son, Gregg, and his second son, Scott, who had recently joined us. We had made the bank an offer to buy out the mortgage at a discount and purchase the building at a price we could manage by putting other people together, along with our own resources.

After arriving at two o'clock, they kept us waiting in a conference room for hours, supposedly because they were calling the other banks. I wasn't feeling well and was so sick and tired that I lay down to rest on the conference table. About seven o'clock, Drake finally came in and told us he'd received answers from the other banks. None of them, including EAB, would accept our offer.

Knowing we had only guaranteed $10 million, and fortunately not the entire $100 million, I took another approach. "We'll walk away and give you the keys, along with seven million of the $10 million."

"Oh no, you won't. I'm going after the hundred million," Drake said coolly. "You'll be getting a foreclosure notice."

We didn't think they had a legal right to do that, but nonetheless, we found ourselves scrambling to defend our position. This began a huge negotiation that lasted for a year and a half, and never seemed to go anywhere, like a dog chasing its tail. It was the most difficult time I had in business in my entire career.

Word leaked out when the bank put our loan into the "bad bank." Rumors, started by another builder, claimed that Reckson was going bankrupt. It was a different builder this time, but our usual Moriarty. At the same time, a large Long Island industrial company that we were friendly with declared bankruptcy, which only added fuel to the rumors. Other than the problems with the Omni Building, our business was going well. We were still actively building, very much solvent, and had few vacancies in any of our office parks.

Calls started to come in from brokers and contractors that we'd been dealing with for twenty years, along with other creditors. Sometimes it doesn't matter what is true—it matters what people perceive as true. It was a matter of time before EAB would serve us a default notice and that would become public. I knew I had to make a statement to allay the panic, so we called a press conference on the day the foreclosure was served on Omni. Two newspapers, two television networks and a radio station showed up.

The main point I had to get across was that even though we might lose Omni we were definitely not going bankrupt. No matter how long a story you gave to the press, they would take just one "sound bite" out of it. Knowing I had to make it good, I stepped up to the microphone and confidently said, "I've got good news and bad news. The bad news is we *may* lose Omni. The good news is we have seventy other buildings." That was the headline to every story, and it stopped the rumors right in their tracks.

During the seemingly endless wasted months of bank negotiations, at some point I realized we were sapping our resources and putting time into Omni that should have been dedicated to the rest of our real estate. We *did* have seventy other buildings. I wondered if the best thing might be to stop fighting and let Omni go, even though I knew the boys were still keen to fight.

I agonized over it all night. The next morning, still undecided, I looked at myself in the mirror as I washed up. "We are losing a legacy," I said aloud to the sad, bleary-eyed face staring back at me. It was the only time in my life I ever remembered feeling sorry for myself.

"Schmuck," I said to my reflection, realizing that I done the very thing my father had warned me against. He'd told me never to fall in love with a building. We were losing an inanimate granite object, not a legacy. Our sons were our legacies. It was not long after this realization that the tide turned in our favor at the bank.

After our meeting with Drake, when I was in Florida, I got a call from the boys who were upset. We had signed an agreement with the bank that during our negotiation period, the bank couldn't partner or try to sell the building as long as we turned over the management and cooperated with them. We upheld our end of the bargain, and there was nothing preventing us from looking for partners. Scott and Mitchell had visited a large Wall Street investment company that invested in real estate, and when they told the broker about Omni, he pulled a thick proposal from his drawer and asked, "Is this the building you are talking about?"

The bank had centered their proposal around our building. They already had someone in place for a small percentage and now were trying to find a larger partner. It looked as if they were trying to set up another builder in business. When the boys told me about this, I realized we were being played for suckers. The bank had no recourse because we were totally cooperating.

I told the boys to leave it with me and not do anything. I hung up the phone, and sat down to think it through, wondering for the first time if it was all phony. What if they never intended to make a deal? What if they had their own people all along? What if this Drake guy is getting paid off? My questions weren't all that far fetched because I'd seen so many ridiculous things already, and now I was putting two and two together. They had been stalling the whole time.

Looking back now, I could see their reasons never really made sense. All through these going-nowhere-negotiations, if we offered something, they would hedge. They would use ridiculous excuses like they "weren't sure" or they "wanted to have their lawyers look at it." We never moved forward and never came to any agreements about anything. A seasoned negotiator, I always gave as much as I took and I'd never had trouble getting to the bottom of a deal before, but this one was impossible. I started to think that maybe, from the very beginning, they'd never had any intention of negotiating to a conclusion with us. *What if they'd never even called the other banks and had never given our original offer to anyone?*

I knew two of the bankers from the other banks, so I called them up and asked, "A year and a half ago, when this started, were you called and given an offer from us?"

They both told me, absolutely not, their banks were never given an offer.

When I explained what had happened, they started to talk to me. I found out the other banks were getting as much of a runaround from EAB as we were. I sent the banks a letter explaining the offer I had made eighteen months earlier. Dime Savings Bank and Union Bank were extremely angry, as were the others.

I was furious too, and accused EAB of breaking their contract by trying to sell the building. Drake denied it, but we got copies of the paperwork with their name on it. After I showed him the proof, he blamed it on their managing agent. Regardless, it was their responsibility.

The boys were upset about the foreclosure and wanted to fight it. We hired Weil Gotshaw, the top bankruptcy attorneys, to help us show the bank we were in for the long run. We sent an official offer in writing back to EAB. It was exactly the same offer we had made in the beginning, but this time we copied all the member banks. Our offer was $64 million for a $94 million dollar loan, plus $10 million in interest. Our deal was that we would put $10 million dollars in escrow and the keys to the building, which meant they would own it and keep our $10 million.

We had ninety days to come up with another $54 million dollars. When we signed the deal, EAB never thought we would be able to raise the money in that climate within ninety days. They still thought they had us.

Mitchell knew one of the fellows at Odyssey Partners, the same Wall Street investment firm the bank had sent the proposal to. The investors came out to the building and interviewed us, and then we went back and forth negotiating a price. It was not the first time this happened to me and wouldn't be the last, but I worked out the final agreement with Odyssey in the bathroom with one of the principals. When we walked out we had a handshake. (We washed first).

Odyssey put in 16 million dollars for 60% of the deal and Reckson put up $10 million for 40% of the deal. A German bank lent the balance. We put our $10 million back in reserve and the other banks accepted our offer, so we bought back the mortgage at a discount, which had been our original plan. We had made this same offer two years prior, and for all that time we weren't carrying the mortgage, weren't making payments. Even though it took a tremendous effort and a lot out of me, it worked to our benefit in the end. They say a diamond is a stone made good under pressure . . . any more pressure, it would have been a flawed diamond.

For almost two years, Scott and I had gone to the bank almost every week to negotiate this deal. Scott was twenty four years old at the time, so it was quite a lesson for him to be involved in this. One of the more comical parts of the story happened right at the end when we were coming to the final agreement. I was driving into the city to the bank and I had a bad cold. It was summertime and the air conditioning wasn't working in the car so I had the windows open. I was on the phone to Scott when we drove past LaGuardia Airport and I got a whiff of the jet fumes. All of a sudden, I lost my voice—it wasn't just raspy—I had no voice at all and I was on my way to the bank to negotiate the final, critical part of the deal.

I couldn't tell Scott over the phone that I was unable to talk, so he thought the phone had disconnected. When I met him in the lobby of the bank building, I motioned to my throat and shook my head. As we walked into the office area, I leaned over and whispered in his ear, "Good luck. You're flying solo." Although I had done the talking up to this point, Scott knew the presentation because we worked on it together. He did a great job with the final close that day, and he hasn't stopped to look back.

Later, several people said, "How old is he? Thirty-two? Thirty? I can't believe anyone that young could be that professional and that articulate." When I told them he was twenty four, they almost fell on the floor. By the time the Omni escapade was over, our sons had each earned their "Doctorate in Advanced Real Estate."

CHAPTER 18

Terrapin, Travel and Toddlers

In 1989, we took our first family vacation where all the boys could take their spouses or fiancées. We went on a large sailboat, *Wind Spirit*, and sailed around the Caribbean. Debbie was pregnant at that time and Willi was born after we returned on March 28, 1990. Our first granddaughter, Willi came out of the womb as a young lady and continues to be a beautiful person. Since Judy and I never had a daughter, Willi, who was named after my father, holds a special place in our hearts.

We'd lived in Dix Hills for 25 years and by the end of the eighties all the boys had left the nest and weren't coming out to suburbia to visit very often. Having always wanted a large piece of land to build on, I had been looking for a large, estate-like piece of land for two years. The developers liked to break up the old estates in order to build housing developments, and the estates I'd seen so far were either chopped up into smaller lots or they had a big, old mansion on a long, narrow piece of land. I had just about given up looking for a piece of land on which to build a contemporary house.

Knowing what I was looking for, a broker I knew called me one day to say that a very exciting property, Terrapin Hill, had just come on the market. It was a house she didn't even have to look at, she told me, because she used to play there when she was a child and knew it well. She made an appointment with the owner for us to view the property right away.

My heart jumped as I drove up the long driveway, lined with colorful rhododendrons and large trees. I knew I was going to buy this house before I had ever seen it. The driveway curved gently as it wound its way to a large courtyard, with a center fountain and an exciting view of the front of the manor house. The driveway and 14-acre property was exciting because it had been designed and landscaped by Frederick Law Olmsted, the father of American landscape architecture. He had planned Central Park and Prospect Park and many major landscaping projects throughout the northeast and the rest of the country. The owner didn't even have me go through the house at first, but met me at the far end of the beautiful 300-foot back lawn. He knew he was selling a stunningly landscaped piece of land, not just a house.

When I turned around finally to look at the house, I saw a lovely brick Normandy-style manor house, straight off a postcard of a French estate. I'd been inside these old houses before and they never flowed well; you had to walk through one room to get to another, and the next room to get to the next. You could never install air conditioning properly or a sufficient electrical system because in order to get into the walls and ceilings you would have to knock down the lovely old plaster walls and destroy the house. I had little interest in looking at the house because I assumed I would build a new one. Besides, I had every intention of living in a contemporary house.

When the owner said, "Come take a look," I didn't want to tell him I was already sold on the property alone. I was amazed when we went inside. It flowed well and was *very* livable, unlike any old house I'd ever seen. It had been well-designed architecturally as a country home with a more open feel and with a lot of light, and it didn't have all the extra gingerbread that most of them had. The owner had bought it five years earlier from the original family who'd lived there for about sixty years (originally part of a 130 acre estate), with the idea that he and his wife would live in it part-time and do a handyman decorating job on it. Rather than spending money to restore the elegant wood walls or floors they had slapped deck paint on them. I didn't know if anything could be done about those, but there was no way I could live in a home without air conditioning, and with our sizable art collection, we needed extra electricity. So I figured I'd have to knock down the house anyway. Nevertheless, I made an immediate offer, even before Judy saw it. I always felt procrastination was the mother of failure.

The house was so beautiful that I felt it was worth finding out whether it could be restored. As I got into it, I uncovered the history of both the house and Sydney Mitchell, the original owner, one of three founders of the exclusive Piping Rock Golf Club. Mitchell had started the company that was the predecessor of Con Edison, which built the major power plants that served the northeast with electricity. Until 1933, houses didn't have electricity or even conduits in the walls in preparation for it because they didn't know electricity was coming. This house, designed by Connors, a famous New York City society architect, had been started in 1924 and finished in 1927. It was built like a brick fort with great structure and magnificent workmanship. The forward-thinking Mitchell had put conduits in the walls because he realized what was happening with the advent of electricity. Not only did he have conduits in the walls, but he'd left plenty of space between the first and second floor where air conditioning ducts could be put in. The man was way ahead of his time.

We were able to restore the house, install everything we needed and restore all that rich wood. I ran the job myself with my own people. The work was extensive, but it took only six months from the time we started until the time it was totally furnished and ready. We moved in May, 1990. Compared to Dix Hills, a five minute drive from my office, Terrapin was about a twenty-five minute drive on the expressway, but I didn't mind the extra commute. To this day, I pinch myself every time I drive up that driveway. *"Success is getting what you want. Happiness is enjoying what you get." —Donaldism.*

Once we moved to Terrapin we had a lot more visitors, even sleepovers with the grandchildren on rare occasions. Our granddaughter, Willi, had arrived on the scene about the same time as our new home, so we had her baby naming at Terrapin. We now held major events at Terrapin with as many as 300 guests. We carried on our famous Fourth of July parties and held Mark and Jackie's wedding there, as well as charity affairs and charity art tours every year. There were two swimming pools, flower gardens, a couple of fields, playgrounds and tennis court. The fourteen acres were so spread out that it looked like there was half the amount of people until the dinner bell rang and all 300 would sit down for the barbecue. On more than one occasion the caterers didn't believe Judy and me and because everyone was dispersed all over the place, they were caught short of prepared food. They hadn't cooked everything. We lived in the whole house, and enjoyed having our ever-expanding circle of family and friends visit.

A large house and grounds like Terrapin required help. In the period we've lived there, we have been fortunate to find two great "house couples" to manage the house. The last ones, Baldo and Alex, retired a year before this writing. They were like family to us. A Jack-of-all-trades, there was nothing Baldo couldn't do: from building a building, to masonry, to barbecuing. Alex was sweet and a terrific cook.

The first day that I saw Terrapin, I met Clifford Grandison. He was a fine gentleman from Barbados, about five or six years my senior, who had been caretaker of the house and grounds since 1965. Not only had he lived on the grounds and had known the original owner, Sydney Mitchell's wife, he had worked for the son when he'd lived there in the 1960s and 1970s. He'd stayed with the house when it was basically abandoned and rented to hippie families, then remained through the last owner's five years. Today, Clifford grows the greenest grass, has been a friend, the estate historian, and a joy to see every morning as I enjoy a laugh with him.

Another good friend who's been driving for us for over twenty years is Alton Kinsey. More than once I've been caught short or have forgotten my wallet and Alton has bailed me out of the best restaurants in New York. For the most part he sleeps while he waits for us, but there have been occasions it's been hard to wake him. The other good news is that Alton only sleeps when the car is parked.

The year after we moved into Terrapin, Mark married Jackie Gewirtz on August 1, 1991. We had a large tent in the middle of the gardens and held a lovely, fun wedding for about 300 guests. As a surprise for his bride, Mark wanted to have doves fly during a certain part of the ceremony, so he had Frank Kramer, a real outdoorsman who worked for Reckson for many years, hiding in the bushes near the gazebo with the birds. While the vows were being spoken I could hear this weird rustling in the bushes and wondered if it was the fox we'd seen earlier. Finally, at the end of the ceremony the doves flew out in a mass of white, beating their wings like crazy. Not only did they leave behind some gifts for the guests, they were disoriented in their flight, but it *was* a romantic gesture.

In the nineties we decided we had been going to the South of France for long enough and now it was time to travel our own country. We went to Aspen, Jackson Hole, Arizona, New Mexico, New Orleans, and to Napa Valley and other parts of California.

In 1992, we went to Oregon with Madi and Bob Kaplan. We drove all over the entire state, staying in different areas, and discovered Oregon was one of the most beautiful parts of the country we'd seen. When we got to the Columbia River, we stayed at Ocean Beach, right on the water. One of the reasons we stayed there was because they were supposed to have great salmon fishing; you could fish from a boat in the bay just where the mouth of the river met the ocean. We learned we could hire a rowboat with an outboard motor and a guide who supplied everything at the dock in front of the hotel, but we were told to get up early because it was busy. I said, "Bob, we've got to get up real early and get out there. They tell me the fish bite just for a little while."

We arranged for a six o'clock wake-up call, figuring we'd be on the dock by six-thirty. I got up quietly and crept out of the bedroom so I wouldn't disturb Judy. In the sitting room area, I opened the curtains and, to my amazement, saw about 200 rowboats in the bay already trolling back and forth. Bob and I dressed in a hurry and ran downstairs, managing to get a boat and a guide. At the dock, the guide gave us instructions on how to use the rods, and as he started to pull away, he told me to drop a line. We were about 20 feet from the dock and the minute I dropped the line it pulled. It felt like I'd hooked the dock. "I think I'm stuck," I said to the guide, somewhat embarrassed.

"Maybe you caught a boot," he suggested. He came over and felt the rod for me. "Okay, start reeling," he said.

I pulled in a 25-pound salmon, and I had only been in the boat for about a minute. Amazed at my luck, we had the guide turn the boat around, and Bob and I rushed upstairs to show the salmon to the girls. They didn't believe it at first, and we had to convince them we hadn't gone out and bought it from someone. That morning I was told no one had caught anything yet. Not only was it a hilarious story, we had the salmon smoked and it was excellent.

Mitchell and Debbie's second child was born on December 15, 1992, a son, Benjie. We served the smoked salmon that I'd caught at his baby naming, his bris, held at Terrapin. I don't know which I was more proud of, my grandson or the salmon. Well, of course, I know, but I had a great time telling the big fish story.

On August 24, 1993, Glenn married Tracey Weaver in the beautiful Burden Mansion on the Upper East Side of Manhattan. Tracey had known Glenn since they'd attended Ithaca together. They then spent two years in Michigan and then several years living near each other in Manhattan, going together for all of twelve years. They made a wonderful couple, similar in many ways, both articulate, bright, political and creative. She had an older married sister, Kim, and a younger brother, Clark. Her parents, Jan and Clark, still live in Strongville, Ohio, where Tracey was brought up.

For my sixtieth birthday in 1994 we took thirty eight friends and family members to the Hyatt Gainey Ranch in Scottsdale, Arizona. A couple of months earlier, Judy and I had flown out to Arizona and planned this exciting four-day weekend where we stayed in the hotel and arranged dinners and events. Our guests traveled at their own expense, and most extended the trip with some other vacation in the southwest.

That weekend, we had a full schedule. We had barbecues in the evenings and sunset rides across the desert in Jeeps and horseback riding in the desert during the day. We played golf and had a fiesta where everyone dressed in western clothes. It was a lot of fun for everyone and a very special birthday for me, since I was a closet cowboy.

> *"As you get older, first you forget names, then you forget where you put things, then you forget faces, next you forget to close your fly, finally you forget to open your fly."* — *Donaldism.*

By 1995, it seemed as if we were having a grandchild every year. Jackie and Mark's first child was born March 20, 1995. Jackie gave birth to Jade by natural childbirth at a birthing center. With Mark at her side, she put herself into a meditative trance without taking any shots or drugs. Reportedly, Jade came into the world bowing. She is such a little diva who is always "on."

When our first grandchild, Willi, was born, we went to the hospital. Men of my generation did *not* go into the delivery room, so I never had the experience that Mitchell, and now Mark have had. I never even changed a diaper, much less helped deliver a baby. I must admit that when I was at the hospital and saw how absolutely euphoric Mitchell was–how he was laughing and crying at the birth of his daughter–I realized that I had missed out on an important experience. I felt somewhat envious of my sons.

About the same time Jackie was giving birth to Jade, Mitchell bought two acres to build on in Brookville, just ten minutes from Terrapin. He designed the house with Mark. While it was being built, Mitch and Debbie sold their condo in the city and moved with their two children, Willi and Benjie into our house for eight months.

We had a good time with the kids during that period. I remember Benjie getting easily frustrated. At the age of two and a half, we had his eyes checked and it was discovered that he needed pretty strong glasses. When Debbie put the glasses on him, Benjie took them off and threw them against the wall. I said to Debbie, "How do you expect a two-year-old ever to wear glasses?"

She calmly picked up the glasses and put them on him again. This time he looked across the room and started naming things on the wall and toys that were in shelves. It was as if he were seeing for the very first time. From that moment, he never took his glasses off again. Feeling bad for him, I decided to join him and started wearing my reading glasses. As he grew older his vision improved some and he got contacts. He is eleven at the time of this writing, and even though Benjie got rid of his glasses, I'm still wearing mine.

The next year brought another grandchild. Milo was born to Tracey and Glenn on December 14, 1996. We had stayed home from Florida that Thanksgiving to await Milo's birth and that's when I stumbled across my father's paintings at the auction that I told about earlier in this book.

Jackie and Mark had their second child, Dylan, born August 1, 1998. Then Tracey and Glen had twins, Sasha and Ruby, born April 14, 2000. As each grandchild came, they made Judy and me ecstatic.

By 1995, we started taking the entire family on week-long vacations, grandchildren and all. We would rent beach houses, mostly on St. Bart's, which was our favorite, but we also went to Barbados and Anguilla. The houses were always within a short distance of a beach, so the kids could play and do water sports. As the family grew, it became harder to find a good house that could accommodate us all, so we sometimes had to split up and get two houses near each other, also not easy to find. In the end, there were eight adults and seven children. All the cousins loved being together and kept each other busy and entertained. About three of the nights during our one-week stay we would all go out for dinner and the rest of the time, we had chefs come in. There were some very good French chefs on St. Bart's. We did this for about eight years, then as my sons' children got older and busier, the timing was harder to coordinate and they started taking different vacations on their own.

A few days after our thirty-ninth anniversary Judy and I dressed in our formal attire and joined Bev and Bob Yaffe to attend a charity function at the Craft Museum, located across from MoMA on 53rd Street. About six weeks prior we had received our invitation. It was a beautiful museum, one that we had been active in as collectors. When we arrived, we were taken up to the second floor. As the elevator opened about sixty people screamed, "Surprise!" I thought it was Bob's birthday or something that I had forgotten. Then people started to congratulate us on our fortieth anniversary, even though it was our thirty ninth! Our three sons and their wives had given us this wonderful party in that beautiful museum. It was probably the first and only time in my life that I was truly surprised.

A year later, we celebrated our fortieth anniversary in 1997 with Stanley and Iris Rabinowitz, and Bruce and Peggy Tucker, who were all married the same year. We had been celebrating other anniversaries together all along the way. Since this was such a big one for all of us, we took a three-week vacation and cruised Australia and Bali. We snorkeled on the Great Barrier Reef and then explored the reef, going down in tanks like little submarines so we didn't need to scuba dive. We also rode out a cyclone on the cruise ship. The cabin Judy and I had was in the bow of the ship and we were rocking up and down in over 40-foot swells. It's hard to believe, but I found it restful and slept through most of the storm. The only problem was when I had to get up to the bathroom because walking was tricky.

The cruise concluded in Bali and we spent over a full week there. This was our second visit to romantic Bali and we stayed in two different Aman hotels, plus a Four Seasons on the beach where each villa had its own private pool. It was as close to being in paradise as you could get.

Bali is naturally magnificent, historically artistic, and culturally beautiful. We enjoyed the wonderful boutique hotels, rated the finest in the world, and seeing the sights. We had a great guide and she took us up in the mountains where we purchased some art and furniture that we now have in our house in Boca. We often went out alone without the guide, sometimes as a group. Usually we all went sightseeing together and to the entertainment at night, like Balinese dances and such. I'm glad we were able to do it then, but I feel sad for our children and grandchildren because they probably won't as Indonesia is such a difficult place to travel to today. Bali is a Hindu area and not a problem, but it is part of Indonesia which is actually the largest Muslim country in the world, as strange as that seems, and there recently has been too much trouble and terrorism there for it to be safe for travelers. With the sad realization that we probably would never return, we remodeled our Mahogany Bend home in Boca Raton after Bali.

CHAPTER 19

Going Public
and Internet Mania

After we got Omni back, we owned 40% with the right to buy the other 60%. The economy was still bad in 1993 and 1994. As I explained, the recession had hit Long Island particularly hard. We managed to keep our key people; many had been with us for 20 years by that time, and the people we kept were all loyal, bright and hardworking individuals like Norman Berlin, Muriel Klopsis, Kathy Giamo, Eddie D'Orazio and our secret weapon, Harry Stavro.

Originally from Greece, Harry started with Reckson as a maintenance man. Over his twenty years with us, he worked his way up to managing all our industrial business; he ran our maintenance crews and leased the space. Whatever the problem, we could send Harry in and he would fix it. The tenants loved him and he kept everyone happy.

Norman Berlin, my brother-in-law, did very well after he left Wall Street, heading up our industrial leasing department. Norm had his own way of doing things, would worry over deals and at times would sit and watch, waiting for the phone to ring after turning down an offer. He still collected Wall Street Journals and had several weeks worth of them on his desk. After moving with us to the public company, he retired six years ago after twenty three years.

Kathy Giamo headed up all our office rentals and rented Omni during the difficult years. It was unheard of having women executives in real estate at the time we hired her. She was very bright, knew Long Island well, and belonged to all the industry organizations. She left Reckson to start her own business shortly after four of the Rechlers left to go into private business.

Muriel Klopsis started as Roger's secretary, then became his executive assistant and then ran our entire Customer Relations/Property Management department until she retired about 2001.

During the eight years that I was with the public company, I had turned over the creative marketing and design and annual reports to Carol Allen. For the previous thirty years, I had never felt comfortable with anyone else doing it, so I had always done that job myself. Carol was talented, bright, and a pleasure to work with.

Martha Yaccarino was originally with VIP. When Reckson bought the industrial portfolio she came with it as administrator. Dedicated and bright, Martha continues to work with us as of the writing of this book.

The "other women" in my life were my three executive assistants: Beverly Thompson, Loretta Regrutto, and Patty Fleishman. Beverly started with us in Bohemia, became our office manager and bookkeeper and finally left me after 15 years when she married and moved to Miami. Loretta followed Beverly, and although sharp and biting, people liked her. She retired after about 12 years with me. For the past five years, Patty has been there for me, coming aboard in the public company and then moving with us to the private sector. She's great on the phone, has a terrific personality, and helped with this book. Without the help of these three women over the years, I would have had trouble finding a pen, much less remembering where I was going.

> "Behind every dark cloud there's always sun. You
> have to make sure you can hang around long enough
> for it to come out." —Donaldism.

Grumman, Fairchild and Republic Aviation had been big employers of Long Islanders. All of them aviation companies, Grumman alone used to employ 26,000 people, and by the early nineties, they were ready to pull out with their last 8,000 employees. Part of that pullout was on Route 110, the Melville area where the bulk of our office space was. Grumman left 1,300,000 feet totally vacant which forced a downward pressure on rents. All that empty space put prospective tenants in a prime bargaining position, and landlords now had to make concessions, such as offer free rent for the first year and/or expensive office renovations.

It seemed as if the tenants all wanted gold on their floors. Our capital investments became very expensive while our rental income dropped by 30%. At the time, the banks were buried in bad loans and stopped lending on all real estate. Luckily, our existing mortgages were very conservative, so even with the rent reductions, we were able to operate. I was confident the economy would rebound because Long Island had a very special entrepreneurship and a highly educated work force.

As I mentioned before, in the coming years, many of the people who lost their jobs with Grumman started their own businesses and were successful. The Island changed. Within a two year period, the Island shifted from a defense-oriented manufacturing economy to a diverse service-oriented, hi-tech economy. Our empty office space on Route 110 was filled by companies dealing in financial services, insurance, health care, and law firms.

From 1991 through 1994, Reckson devoted all its considerable resources to keeping our tenants and adding new ones to fill up the bankrupt spaces in our buildings, as well as finding new tenants for Omni. Although there was a lot of good real estate available at great buys during this time, we weren't in a position to take advantage of it because of the capital outlay, our diminished rental income, and the $10 million we had put into Omni.

There was no conventional financing available, and it appeared that the banks would not be making real estate loans in the near future. In the economic climate of 1994 we began to look into going public as a Real Estate Investment Trust (REIT).

I had never been an enthusiast of the stock market—after all, I was my father's son—but I considered taking the company public. As I looked into it, friends and tenants involved in the public markets, such as Bernie Madoff, advised me to stay private. Bernie felt Reckson was too entrepreneurial for the market. I came to the same conclusion, perhaps partly because I was still vulnerable from the Omni debacle, and told the boys—our sons—I didn't feel it was a good idea to go public. "If we don't go public, we'll have to be landlords all our lives," they told me. "We want the same opportunity to build and create like you and Roger did."

Well, I couldn't argue with that.

By 1995, we started looking at and talking to various Wall Street bankers. To become public you have to register on the stock market, then go out on road shows to promote yourself. Basically, the bankers take you out and sell you to the public and people buy stock in your company. It was a tough time for a real estate company to go public because the excitement of Real Estate Investment Trust IPOs, or Initial Public Offerings, had come to a halt. (IPOs are companies coming onto to the market to be offered for sale.)

Scott was able to put together a group of Wall Street bankers headed by Steve Cantor from Paine Webber. Steve was a Long Island guy whom I liked a lot, and he understood the hardest sell would be that we were a Long Island real estate company. Throughout the rest of the country, people saw Long Island as a bedroom community to New York City, not a business community. Paine Webber was to be the lead, and the other Wall Street banks included First Boston, DLJ, and Alex Brown. We were all ready to go when Paine Webber merged with another company, Kidder, and replaced Steve Cantor and their entire real estate department with Kidder's Real Estate Department. Cantor and his group went to a smaller boutique firm that couldn't be the lead. We didn't like the Kidder group that took over and they didn't like us.

I wondered if we had developed a pattern. No big deal went unpunished. These kind of deals were never easy for us, and going public proved no different. When it became obvious that the new people were only interested in selling us to another real estate investment trust rather than taking us public, we scrambled. First Boston took the lead, but they'd had very little experience as such. We stayed with them a number of months rehearsing, rehearsing, and rehearsing. Nothing was happening except that I felt like a puppet and was getting depressed.

One day, I had lunch with Lew Ranieri, a tenant who had become a good friend, and he counseled me on going public. Lew, by this time, was a legend on Wall Street. In the eighties, he went from the mailroom to the number two man at Solomon. He created securitized mortgages, and has an uncanny feel for the market and the economy.

"Forget what they are trying to turn you into," Lew said. "Just be yourself. The market is going to love you. You, Don Rechler, are an easy sell. You have a down home philosophy and a sincerity they are going to eat up. You can learn what they have, but they can't learn what you have."

It was then evident to me that, whatever you're trying to sell, the most important sell is yourself. From then on, I did it my own way and was myself.

Not long after that, it became clear that First Boston wasn't going to be able to take us public. When we pushed them on it, they admitted as much. We were scrambling again. That same day, we called our attorneys, Brown & Wood, who were also the attorneys for Merrill Lynch, the early leader in the REIT industries. Richard Saltzman from Merrill Lynch had started the whole new REIT craze with Kimco. We met with him and hit it off well. Their analyst, Jordan Heller, loved the family story even though most of the market traditionally didn't like mixing family and business.

Merrill Lynch got ready quickly to take us on a road show. We were going to be the first REIT IPO in nine months. Our first stop was local, in New York City. I had trepidations about how we would do because I really thought I didn't know anyone in New York. By a stroke of luck, it turned out that in four of the first five meetings, we found a connection. One was the nephew of a fellow member of a club I belonged to, the Glen

Oak's Country Club. I had done business with another man's father in real estate, a man who had thought highly of me. Another from Long Island had used the gym at our Omni building and seen the quality of our building work. Reputation made for an easy sell.

That lucky day, I happened to be wearing two watches because I got one back that was out being repaired. From that time on, I never took the two watches off.

Merrill Lynch had felt Reckson would be a difficult sell because of three factors: it was a family business; it was only on Long Island; and timing was poor. Scott and I flew around the country for ten days by private charter at our expense, meeting with institutional buyers. We stopped in as many as four or five large cities a day for meetings. Most meetings lasted an hour, except lunch and dinner meetings where we would do slide shows for several investors at a time.

Meanwhile, Mitchell, Gregg and Roger hit all the local retail brokers who sold to individual investors, even though historically, there wasn't much retail buying on REITs. They were usually 90% institutionally owned. For instance, Fidelity Mutual Fund would buy them—big, big companies rather than individual people. However, we felt we would do really well on Long Island with retail investors. After all, Long Island is the equivalent to the fifth or sixth largest city in the country, and Islanders tended to support their own.

Outside of New York, the sell was much tougher. We told them about the Island and what a terrific workforce we had. We explained we were in an area where it was very hard for anyone to compete with us because of what I called "dukedoms and dutchdoms"—we had 90 separate communities to deal with, which meant 90 separate local governments and taxing authorities.

As it became clear that it was going to be difficult to raise the necessary money, the Merrill Lynch people suggested that we take $60 million worth of real estate out of the offering and reduce it to $162 million instead of $225 million. We gave the REIT options on the properties we took out; they could buy them at a later date.

As a family, we had committed to buy an additional $10 million of the stock ourselves, as individual retail stock-buyers. With that, and the *amazing* amount—about 35% retail sold mostly on Long Island (because so many people knew us)—it was still touch and go whether or not we could raise the $162 million we needed.

On the last day before we went public, we were all gathered in the Merrill Lynch office in New York, waiting for the market to close. It was like an election night waiting for the outer returns to come in. Just before closing, the Merrill people were actively working the phones, and two regular buyers raised their orders. (The institutions that heavily bought these REIT stocks were called *REIT Mafia*.)

At the close, we were still five million dollars short.

Merrill had a meeting and decided to buy it themselves. Going public usually takes six to eight weeks, but exactly nine months after conception, Reckson Associates Realty Corporation, a public company on the New York stock exchange, was born May 26, 1995. It had been a difficult pregnancy and a hard birth.

An interesting side story involved applying for our call letters, the letters we would be recognized by on the stock exchange. We were told we would probably need to go up to four letters. I really wanted RA, our logo, but it was hard to get two letters because there were so many used. "Don't even put it down because it's going to be tough to get three, no less two," everyone told me. We were only allowed four choices, then we would have to resubmit. Undaunted, I listed RA as our first choice.

Luck was with me again and we got RA! Republic Aviation, the company that had previously used those call letters had been located less than four minutes away from us, had been out of business for five years.

RA was off and running. We put together an outstanding independent board of directors: Lew Ranieri, who ran Hyperion Funds; Len Feinstein, co-C.E.O. of Bed, Bath & Beyond; Harvey Blau, Chairman of Griffin, a hi-tech company; and John Klein, Managing Partner of one of Long Island's largest law firms and a former Suffolk County executive who had known my father. We rounded out our board with a retired banker I had known for years, Connie Stevens, and Professor Herve Kevenides who was French. All were dedicated, bright people, who contributed a great deal of business knowledge to our company.

Lew and Len were my confidants and I valued their business instincts more than anyone else I had ever met. As a partner and founder of Bed, Bath and Beyond, Len was a keen businessman with exceptional foresight and insight. Equally adept at art, sports and politics, he was a competitor in all. Both of these men were close friends of mine, and Len was particularly social and fun to be with.

Our business strategy was simple: to expand into the other Tri-State area suburban markets using the same formula that made us successful on Long Island. Basically, the formula consisted of five main points.

1. We believe real estate is a *local business*—to be successful you had to know the local market and be involved with the local communities.

2. We are a *full service* company with leasing, property management, architecture, construction, development and finance, all in-house.

3. *Dominate* whatever market you choose. Pick the best sub-markets and concentrate on them. Try to control one third to fifty percent of the market until you are the largest, or at the very least, the second largest landlord of that sub-market.

4. *Create value*. Recognize and buy underutilized property. Create significant value by repositioning or redeveloping such as we had done in North Shore Atrium, and try to deal infill markets—places where land is scarce, where the property is already developed.

5. *Franchise value*. About that I have a saying: "Do the right thing—to all constituencies—tenants, brokers, community, employees and the trades."

With our formula as a guide and a market that had been starved for capital, we started buying value-added deals on Long Island and were able to get large returns on investment. By 1996, our first full year in business as a public company, we expanded into Westchester, New Jersey, and Stanford, Connecticut. All within a fifty-mile radius of New York City, we purchased office portfolios with management teams.

That was the easy part. The hard part was to have them buy into our formula and our company culture . . . "Do the right thing," as simple as it was. We hired Franklin Planners, organizational designers out of Salt Lake City, to help us integrate the management teams and merge the companies. We put them up in a hotel in Melville, and they worked with our management team and ourselves for two weeks with intense planning and training.

Having majored in college in a similar field, I was both impressed by the people and skeptical about the concept. I told Scott that this was all well and good, but in my experience, about three months after the planners returned to Utah, 98% of what we had learned would be forgotten and dropped. Two weeks later, I walked through the office and there was Eric, the main planner from Salt Lake City. Scott had hired him, bought him a condo and moved his family to Melville. The man who had implemented our plan now worked permanently for Reckson, so he was able to keep it alive and adapt it as he saw it working. Eric was only one of the outstanding, talented people that Scott, Gregg, Roger, and Mitch added to our company. Some of the other department heads were Carol Allen, Ken Bauer, Tom Carey, Tom Kirwin, Susan McGuire, and Tom Riley.

By the end of two full years of operations, we were hitting on all cylinders. We tripled our market cap, our stock almost doubled from $24 a share to $46, and we built a value creation pipeline of $750 million that would grow to $1.2 billion upon completion, by the end of the following year.

Market conditions started to strengthen. We were praised for our entrepreneurship and real estate savvy, and negatives about family were dropped. This was a pattern that developed with RA and would continue through public life. When business was good, we were touted as sharp real estate operators, but when it was bad, we were a "family business."

In 1997, besides investing in prime office properties in over three markets—Westchester, New Jersey, and Stanford, Connecticut—we invested in three operating companies that we felt could create exceptional future growth for Reckson. One was a big box distribution facility in New Jersey. Eventually we found their company's principals were inflexible and stubborn, so we sold the company, but only after a 16% profit.

We created Reckson Service Industries (RSI) to provide real estate related services to the industry. As these investments were not qualifying income, we couldn't grow these businesses within the REIT because at that time REITs were only allowed to make income from rent. The Securities Exchange Commission treated non-rental income (sales) differently for REITs.

We sold our instant office business, which had grown into a good business, to Reckson Service Industries (RSI) because it was non-rent income and not allowed in the REIT. We then sold RSI to our shareholders. Although the RSI shares were distributed to RA shareholders, it was not part of the REIT and traded as a separate public company. RA provided Reckson Service Industries with a separate $100 million credit facility, a loan with liability limited to that amount.

Our third company was RSVP, Reckson Strategic Venture Partners. The point of this company was to invest in early real estate platforms outside of Reckson's core businesses of office and industrial buildings. RA would use this company as a research and development vehicle for future investments. We committed $100 million as seed capital, along with another $200 million that the George Soros Fund invested for preferred equity. We were able to get two highly qualified real estate executives with successful investment experience to run RSVP.

One was an attorney we knew who sourced deals for Paine Webber's funds, and the other, whom we also knew, was a C.P.A., the managing partner of Ernst & Young's Real Estate Department. We thought we couldn't find two better professionals.

But we were wrong. They were fish out of water in our entrepreneurial environment. They had both worked in very structured, institutional venues, and the entrepreneur part was very difficult for them. We started them out with two deals we had already sourced. One was a student housing business that looked very promising, that we thought could eventually be brought into the REIT. The other was a land deal for future development in the Catskills with the purchase of both the Concord Hotel's and the Grossinger Hotel's land. After those two deals, everything else they did, as a whole, was unimpressive. They didn't know how to make business decisions that would affect them personally. It was like putting two kids in a candy store.

In 1998, Reckson experienced continued growth in all four of our markets. We started to dominate the key sub-markets such as Short Hills, New Jersey, Tarrytown, Westchester and, of course, Route 110 and Mitchell Field on Long Island as well as Hauppauge and Bohemia, our industrial markets. We were now the only landlord with a presence in every one of the Tri-State area's major markets.

It had become obvious to us that Manhattan was the engine that drove Westchester, New Jersey and Stanford markets. Our next move was something I never anticipated: expansion into Manhattan.

It was a difficult market to enter. The major players were New York City real estate families that were established for years. We considered a partnership with a couple of them. After exploring different ways for about one and a half years, it became apparent that wasn't the way to go. Because of the entrepreneurship, they would never be compatible with the public market.

We had an opportunity to buy a billion dollar troubled New York City REIT whose organization was poor. We negotiated a partnership with Richard Rainwater of Crescent, the REIT, for the New York City purchase. Halfway through the purchase, Rainwater had their own problems and opted out. We continued alone. Six months later, we hired Tod Waterman as managing director and we rebuilt the company. Tod was highly regarded in New York City real estate circles, and there was no doubt that Manhattan was the hub that turned the wheel.

As we expanded into New York City we quickly realized we couldn't compete with the immense capital pouring in from large financial institutions in almost every country in the world. They would purchase buildings in New York City at a much lower rate of return than we could afford to offer. This meant we needed "deals with hair on them"—problem real estate that others might fear. No conventional investor would be interested in these type of deals until the problems had been stabilized. With our real estate expertise, we could solve the problems and turn the investment around to our benefit.

Six months into New York City, we were brought a deal at 919 Third Avenue. It was 1.4 million feet at a price of $277 million. The building was much larger than any of the total building portfolios we had bought in the different suburban communities. It came out to $200 a square foot. This price was approximately half as much in comparison to other buildings in New York City.

We were able to buy a first mortgage from the Japanese bank that had foreclosed on it, then negotiated for the title with the original owner, instead of having a long, drawn out court battle that could have lasted years. Half the building—800,000 feet—was empty and not bringing in any income, so we re-leased it, restructured the offices and fixed up this half of the building. We sold one of the institutions a 49% interest in the building and made a $100 million profit on that, which then made our return higher than most people were able to get in New York City.

Three months later, in September 1998, we did another very similar deal with 1350 Avenue of The Americas. That was a 540,000 square foot building that we bought for $234 a square foot, which was about a 40% discount in that neighborhood. After that we were recognized as a "player." We took Manhattan! You know what they say . . . "If you can make it there, you can make it anywhere." Because of our Manhattan presence, we started to get even more activity on our suburban office buildings.

Reckson Service Industries (RSI), the company we had wheeled off from Reckson (RA) to each of our shareholders for $1.10, became public in 1998. Scott, Mike Maturo, our C.F.O., and Jason Barnett, our General Counsel, actively ran it. The rest of the family, including me, was on the Board of Directors but not part of management.

Scott hired an executive from G.E. Capital as the C.E.O. of RSI. Granted, I had a bias regarding G.E. Capital because I always felt they had a cutthroat company culture, and I didn't care for the fellow they hired. The RSI offices were set up in the same building as RA's offices at 225 Broad Hollow Road, so we could see what was going on.

RSI successfully bought other instant office companies, including the largest of all, now called HQ. The other company RSI bought was On-Site Access, a telephone company that wired for phones and Internet in office buildings. I felt both of these companies had a very good chance of being successful. What I couldn't understand was with all these companies they continued buying, why wasn't RSI able to keep the key people who ran them?

Just about that time a lot of the tech Internet companies started going public. Scott changed the focus of RSI to buying Internet companies that would service real estate with the intent of bringing them public. Internet IPOs were the flavor of the day on Wall Street.

A couple of months later, I felt something was wrong. There was a real estate employment agency and a number of other small companies, including a concierge service, and Realty IQ, a real estate directory listing all available space for rent. I struggled with the purchases of these companies because they didn't seem to make sense from a business standpoint. None of these companies were profitable nor was I able to imagine them becoming profitable. Maybe, because I was computer illiterate, I didn't think that I understood them.

The very end of May, I went over to the local Schwab office to check on my account. I was both shocked and amazed to find the office packed, computers rimming the room, televisions going and all focused on the Internet stocks. There were UPS men, cops, teachers, mailmen, housewives and everyone but the Indian chief was there, all passing information on the hottest hi-tech stocks—much more like a racetrack. It was hard for me to believe.

One morning in June of 1999, I was taking a shower and had a flash of insight (I get some of my best flashes in the shower). The Internet companies we were buying—the public companies Wall Street had been buying—were being valued for the most part by "eyeballs." This was a term that meant how many times the computer was hit for that site. Therefore, it was not really a business that was being purchased. They (and we) were buying nothing more than an expensive mailing list with no visible way of making money. Even the top analysts on Wall Street had a formula that a certain amount of eyeballs translated into dollars. It didn't make any sense to me.

RSI got caught up with the rest of the market in [Inter] Net Mania, mass hysteria that feeds on the fear of being left behind. People were caught by a fever, by an anxiety to get rich quick. The Net became foremost in the public's minds. Analysts and consultants made it a big business predicting which Internet companies would be successful. With their eyeball theory, Wall Street bankers and brokers touted Net IPOs. Amateurs, day traders and new players flooded into the Net market with new money and created demand. People mortgaged their homes and borrowed money in order to buy Net stock and be in on the boom. Rationality was long gone. Some Net stocks were selling 100 to 200 times their expected profit. Even smart investors were betting on the "Greater Fool Theory"—that there would be someone who would pay more than you did for a product. That price had nothing to do with any real value. Typically, the bubble market collapses after the "Last Fool" buys into the mania market. Everyone else finally realizes risk is much greater than reward and runs (sells). It was in that climate of Net Mania that RSI was operating.

That June, RSI stock continued to climb. When I told the RSI Board about my feelings and suggested we go in another direction, only Roger agreed with me. Paul Amoruso, a longtime friend whose good business sense I valued, remained silent. Everyone else was blinded by our rising stock prices. We were making money in the stock market, albeit only paper profits. As a potential IPO manufacturer, RSI's main business was now trying to create Net IPOs.

That summer, Scott, Jason and Mike moved RSI out of our offices and into RSI's new offices in New York City. After that, they were rarely accessible, and we started to lose touch with what was going on at RSI.

Meanwhile, the C.E.O., formerly of G.E. Capital, had recently hired several other G.E. Capital people for RSI. I felt it quickly getting out of control. A few months later, at the RSI Board meeting in the fall of 1999, Scott announced they wanted to change the name to Front Line Capital Group. I was against it as much for superstition and luck as for business reasons. After all, I was the guy who wore two watches to every big deal and tattered 20-year-old Giants underwear to football games. I felt our stock would drop with a name change. Again, only Roger voted with me.

By November the name was changed officially to Front Line Capital Group (FLCG) by the Board, which they felt repositioned them with the hi-tech companies. It was listed on the NASDAQ exchange. Everyone thought all the Net firms were soaring and could do nothing wrong in the public's eyes. Front Line stock was no different. In less than two years, since the company had started, the stock flew up from $1.10 and then peaked at $67 in early December 1999. The stock started to slip in late December. In February 2000, Front Line Capital had announced it was about to bring its telephone company, On-Site Access, public when it was hit by a large lawsuit for hiring a competitor's executive against contractual agreement. That forced the stock down.

I was a buyer of FLCG stock from the beginning, through March 27, 2000 as it was coming down and when the bubble burst. I was waiting for the Net bubble to burst. I felt strongly that this was a prime example of my father's bag of potatoes theory. I didn't sell any of my stock until mid-July, 2000, when I started to sell to take losses.

By March, there was more FLCG conflict with Reckson than I was comfortable with, so I told Scott I wanted to leave the board of Front Line. There was nothing I could do there, so I left. Gregg soon followed me.

The stock was in the low thirties by March. The Internet bubble burst in April and the stock dropped to $12. Over the next year the whole thing disintegrated. The hi-tech stock crash has been compared by business people to the great tulip bubble depression in the 1600s when people were trading tulip bulbs for $5,000 a bulb. It was just trading on paper; there was no value behind the price.

I thought HQ, now the largest instant office business in the world, would bail Front Line out because they had $110 million cash flow in 2000, and helped to keep Reckson's loans to FLCG safe. The only trouble was that I didn't know that the new C.E.O. who came with HQ had changed the entire business strategy. Having been a retailing C.E.O., he'd decided to centralize everything. These instant offices were really a local business—the managers had to have local contacts and be hands-on in the market to know the needs of the market and the rents of the competition. Instead of leaving the people who rented the space to do the renting from each local office, he centralized the business and took renting to Washington. Now they were dealing all over the world, charging blanket rents and offering non-market-specific services without any feel for the individual markets.

At the same time they changed the philosophy. This had been practically a recession-proof business because you rented small spaces to tenants for a year or two or three, and as they grew they took more space or you moved them to another building, and started the process with another small tenant. The small tenants grew your business as they grew theirs. This C.E.O. had changed the concept at HQ to renting large spaces to large corporations for a month-to-month period. When the recession hit, the first space that would cave in was theirs, because the large tenant would give notice and leave. The value of HQ quickly dissipated. By summer 2001, Front Line stock had gone down to $4. Shortly after, it went bankrupt, as many other Internet-oriented companies had.

I always felt the original concept we had for Reckson Service Industries was sound. Had it not gotten caught up in the Internet craze and instead had been grown in a conventional manner by building on a strong foundation, it would have been a successful business. The biggest shame was that HQ, because of poor management and a bad business plan, killed a really good business.

RA (Reckson) had established a reserve of $163 million because the loans of Frontline and RSVP were in default, even though the performance of RA's core real estate portfolio was outstanding. All five markets were achieving high returns on all of our projects. During Net Mania, RA's stock went down to $18 and that was after the best five years that any real estate company could hope for. Once FLCG defaulted on the loans to Reckson, the market was unforgiving and RA stock suffered accordingly.

When the Front Line stock hit its high point in December 1999, I felt another casualty from the lure of the Internet. Mark made a decision to leave Reckson and start his own music business on the Internet. He sold some of his Front Line stock and gave up his career in architecture. This upset me.

Mark wanted to put different boutique (specialty) music sites onto the Internet, and thought he would get advertising for them. The advertising idea was a major fallacy for most Internet companies. I explained how I felt about this part of the Internet and how most of the businesses I'd seen weren't businesses that would last; I didn't understand how they could actually earn money because you don't make money by giving away free samples. But music was always Mark's true love, and I could understand that part of his desire. Architecture hadn't been making his artistic and creative juices flow for some time. He was a terrific design architect, and I wondered if he was mixing up his hobby with his business. After talking to him once, I realized he had to take his own road as any young man does. He started his Internet company, but it didn't work out.

Knowing music was where he had to go, Mark formed his own band and made a very good record. The music industry has always been extremely difficult to break into, but Mark works with his first love, music. He has his own band and as I write this is currently writing music for different venues, including the Discovery Channel.

CHAPTER 20

Love of Art

Judy and I started collecting art when we were first married. Back in 1957 Aunt Ruth had pushed Judy and me to buy a Picasso woodcut at a charity auction as a fundraiser for the Israel Museum. It was an edition of five, and was hand colored. We paid $350.

From that time on, we collected wherever we traveled whether it was antiques, or artifacts: African, Asian, or Southwestern crafts. Judy knew I loved DuBuffet, especially his L'Hourloupe period, which were his blue, white and red paintings. This was one of his sculptures she gave me for my fortieth birthday.

For a time, I was buying pieces of art in the $12,000 range, then graduated to about $20,000. Some were very nice pieces, but most weren't what you would have in a major collection.

My uncle Jack had a great eye for art. He had painted when he was younger, and he and my aunt had been collectors for many years. Around 1979, we had seen a very large painting by Pat Stier and it was pricey at that time—$45,000. I asked him if he would take a look at it with me. I met him in the city and he told me the painting was very good, then proceeded to tell me, "Don't make the same mistake your aunt and I made when we first started collecting: Instead of collecting the best, we collected all over the map. Buy only very special pieces and that way your collection will always have value."

He really started me on the type of collection I have today. After giving it some serious thought, I decided that I should start with my favorite artist. In my mind all art starts with Picasso. A year or two later, I was walking along Madison Avenue with Bob and Bev Yaffe. I looked in the window of Weintraub Galleries and there I saw a Picasso. I got very excited. I knew it was a late Picasso oil and I really liked it. At that time, people turned their noses up at late Picasso's. They thought because he was old, his work wasn't good anymore. I still thought he was the best, just ahead of everyone else.

I knew the gallery owner vaguely and he had a reputation of being a tough dealer. Wearing jeans, an old sweater and a cowboy hat, I walked inside and asked him the price of the Picasso. He said, "Why do you want to know? You're not going to buy it."

I pulled out a checkbook, and with a pen poised in my hand, I said, "I've been looking for Picasso's all year." I had a rough idea what it should cost and I wasn't going to negotiate because this guy would be out of my class. "I'm going to give you one chance to give me a fair price. If I think it's fair I'll write the check. If I don't, I'm gone; so you better make sure you give me a price you think I'm going to take."

He got his book out and did some calculations, giving me a price that was 10% over what he paid. He showed me his actual invoice. I bought the Picasso for $140,000. This was a major purchase for us, but I felt that it was as good as any investment I could make, even in real estate.

The critics had been saying that Picasso's late paintings weren't good. A couple of years after I bought mine, the MoMA (Museum of Modern Art) put on a show of only Picasso's late paintings. The prices exploded and suddenly the critics changed their minds. They decided that Picasso was the greatest artist of not only the first half of the Twentieth Century, but of the second half as well. Today it has increased several times what it cost. Besides, we love it.

Judy and I loved the 1960s Pop artists too, and we'd bought a Warhol from the Castelli Gallery very early on. We bought a Wessleman from O.K. Harris and a George Segal from the Sidney Janis Gallery. The most exciting Pop art buy we made was a Lichtenstein we bought at auction. I had started going to the art auctions and learning the market. Just as the real estate market, I got a feel for when the prices were good and when they weren't. I was very disciplined about not overpaying, even if I wanted the painting, because I had decided that every piece of art I bought had to fit in the food chain. For example: you start with Picasso on top at market price, then everything else has to fit in its substantive price behind it.

We had put together a nice collection of Pop artists and 1960's artists. However, we still didn't own our favorite 1960s artist, DeKooning, which we had looked at but never been able to afford, because his good pieces were too expensive. The major auctions featured a large number of DeKoonings, so we were excited to see them. I never felt I would be able to bid on any because I had seen the prices listed in the catalogs and witnessed the final bids at other auctions.

Black Monday hit the stock market, October 19, 1987, and we had our tickets for the evening shows at Sotheby's that

Tuesday. Usually the auction was packed with standing room only, but the day after Black Monday the room was half-full. The auction started and I bid on a couple of lesser pieces, quickly realizing how good the prices were.

When the large DeKooning we'd seen was displayed at the front, the colors were beautiful. Judy and I looked at each other and didn't understand what had happened. Then, being a builder, it suddenly dawned on me and I realized we had been looking at it under fluorescent lighting in the other room. In the right lighting this blend of yellows and greens—that had frankly looked like vomit in the other room—was breathtaking.

They announced it was from the estate of Xavier Fourcade, the gallery dealer that handled DeKooning. This painting was actually Xavier Fourcade's own DeKooning that had toured all over the world's museums and it was now up for auction in an estate sale because Fourcade had died. As I watched the proceedings with interest, my gaze was repeatedly drawn back to the painting. The price was still very low and the bidding didn't seem to be going anywhere. Judy leaned over and said, "If you don't buy this, you will never own a DeKooning."

I jumped in and bid on it, eventually buying it at less than a third of the price of its estimated one million-plus value. We were able to buy three other pieces that day, all excellent values. This happened for two reasons: it was Black Monday, and on estate sales they never put an upset price on the bid (an upset price means they won't sell it unless the bid goes over a certain amount). We had been at the right place at the right time—again.

Exactly 13 months after we bought the DeKooning, we were offered nine times what we paid for it by a Japanese collector. I was tempted to sell, but Judy wouldn't let me. As much as I loved the painting, I felt at that point that it had blown its way out of the food chain. We held onto it, and as the market ebbed and flowed, it dipped down to about twice the price we had paid. It has come back up now, but never to the price we were offered by the Japanese collector.

All markets have inefficiencies, whether it's the art market, stock market, real estate market or retail market. If you are a player, you'll have to follow the market, know the value of what you want, be patient, and the hardest part for a collector or investor: *be prepared to act on a moment's notice.* A market inefficiency could be caused by weather, world crisis, death, or almost anything else. In the case of the DeKooning it was Black Monday, death, and fluorescent lighting all coming together—luck, or residue as Will Rechler would say.

We had done so well with the value of art that I thought it would be a good investment for the boys. About 1987, I started an art-buying partnership with my sons. By the late eighties, the prices on contemporary art had started to soar so high that I found myself settling for less important pieces at high prices. Realizing that was a mistake, I eventually culled them from my collection and switched my focus. Instead of buying contemporary art, Judy and I started a collection of crafts, both glass and furniture, much of which we used in Terrapin since we were furnishing it at the time. We bought glass by such artists as Dale Chihuly, William Morris, and Dan Daily, and we bought Wendell Castle, Albert Paley, and Judy McKee furniture, all still underpriced.

During our travels through the states in the mid-eighties, we had visited Santa Fe. We went to Indian Market with Eddie and Susan Blumenfeld. That was where all the best Pueblo

Indian artists participated in a juried show once a year. I loved Native American pottery, and at the time we were building our house in Mahogany Bend in Florida, and knew we wanted to do it in a southwestern design.

In 1978, we had purchased our first piece of Native American pottery in Arizona. We had collected the occasional piece now and then, but it was a different story when we finally went to Santa Fe where the finest works of the top artists were for sale. At Indian Market they opened their booths at five o'clock in the morning. While I never got there at five, I did queue up about six for something I really wanted. Since it was a juried show, I first visited the smaller viewing area where they were judging the pieces, not where they sold them. There I would scope out which pieces were up for a prize, so when I met the artists, I could discuss the purchase of a particular piece. Much of the pottery Judy and I liked ended up winning ribbons. In the two years that we traveled to Santa Fe, I must have bought 30 prize-winning Native American pieces. I later added to those pieces with the original potters from the Pueblo tribes such as Maria Martinez, Margaret Tafoya from the Santa Clara tribe, and Fanny Nampoya from the Hopi tribe. Today, the bulk of our Pueblo pottery collection, about six dozen pieces, is displayed in our home in Florida.

On one of those trips to Santa Fe we visited the Forest Fenn Gallery. Forest Fenn himself was a real character, an ex-O.S.S. man, which was like the C.I.A. during World War II. He had a bunch of old pots on the top shelves and we saw one among them that we especially liked. It turned out to be from the early 1900s, signed Maria Popova. We bought it for $2,200 and took it to Florida. About 15 years later, I was going through a Sotheby's Native American catalog, and saw what looked like an almost identical piece with an explanation that Maria Popova was really Maria Martinez in her early years. At the turn of the century, it likely had been originally sold at a train station for $5 or $10, just as our pot probably was. The Maria Popova piece sold at Sotheby's for $140,000. Sometimes you've just gotta be lucky. We bought the old pot because we liked it, not because it was worth 70 times what we paid for it. The real point is that we collect what we love.

Sometimes our tastes grow or change, but usually I'm very consistent. A couple of years ago, I sold off a number of pieces I had grown tired of or didn't think were good enough. I often look at my collection, appreciating the aesthetics of design and color, because looking at art gives me pleasure. We have many other collections, but most are smaller than the Indian pottery. We started different collections in our travels: oriental antiques; floral baskets from Japan; primitive art, most from Africa; and Picasso ceramics from the South of France.

By the early 1990s, prices had come back down on Twentieth Century art because the economy had slowed. In late November 1993, I had an opportunity to buy a large Joan Mitchell at Christie's at about one tenth the price it is selling for today. I loved it, but I also knew that buying it at that price was a smart economic decision. As an art collector it is best to be a patient, opportunistic buyer, who knows what you want, then try to buy the best example of what you want that you can afford.

Eleven years after I had bought the DeKooning for such a low price at auction, another similar chance came my way. Christie's was about to move from Park Avenue into a brand new, larger space in Rockefeller Center. They announced they were going to open the new location with a blockbuster show: a collection of Magritte paintings owned by Magritte's attorney, who had just died. It was another estate auction, the largest collection of Magrittes ever shown for sale. This was exciting news.

A typical problem in construction, Christie's new building wasn't ready on time. They had this huge collection of Magrittes, plus all the other exceptional art they'd brought in for their grand opening, and now they would have to sell it in their old space, which was way too small. We waited on line to get in for the viewing, and finally started to wind our way through the show. The pictures were hung two and three in a row above each other because they were so short of space.

I had picked out two possibilities in the catalog. For years, I had been waiting for a Magritte to complete my modernist collection, but I'd never found one I felt I could afford. That day, I walked the whole show and didn't see the *one* picture I was especially interested in that I'd marked in my catalog. I wasn't going to get back in the long line to go through again, so I went to the desk to see how I could have missed it. "I want to see this picture," I said pointing it out in the catalog. "I waited, then I went through, but I didn't see it. I don't see how I could have missed it because I was specifically looking for it."

"Oh, I'm sorry," he apologized. "We should have put a note in here. We had so many extra pictures that we had to send even some of the Magrittes over to Christie's East."

Christie's East was an old factory building where they used to sell inexpensive things like baseball cards and old furniture, things that were collectibles but not antiques. They would never send good art over there because they would never get the buyers. On this one occasion they were in dire need of the extra space, so they had. I told Judy I wanted to go to Christie's East, about thirty blocks away, to see this Magritte. We were with Sue and Lenny Feinstein, but by this time they'd had enough of waiting on lines, so they didn't continue on with us.

Unlike the other Christie's location, this one was empty. We were the only people in the place. There were only about four or five Magritte's, but the one I definitely wanted, that I thought I might be able to afford, was there. We casually asked at the desk if they'd had many people through, and they said only four so far. "Judy, we're going to own a Magritte," I said quietly, grinning.

It was a rare opportunity and, as I said, I am an opportunistic collector. I took a very careful look at everything in the building. I soon owned a Magritte for half the price they had sold for at the other location the night before. At the same time, I bought a Jim Dine triptych (three large 7' x 7' paintings put together), which today hangs in my Boca home. I also bought a unique signed Picasso ceramic and a Hans Hofman—all at less than half what their cost should have been because Christie's had made a mistake and caused an inefficiency in the market.

I have four simple rules about collecting. Buy what you love. Make sure the price fits into the food chain. Be patient and watch the timing and market conditions. Buy the best example of the artists' works you can find.

More recently, we have been collecting photography, and even video art that Mitchell has collected and piqued our interest in. Judy and I, to this day, are excited to go to an art fair or on a tour with a museum or—as we have done with the Craft, Guggenheim, and Nassau County museums—on a trip that lasts several days, where you get to see other people's collections and homes all over the world. One of the more interesting times was when we were in Japan. We were invited to a large home and had to sit on the floor as they served sushi, one piece at a time, as they slid across the wooden floor on their knees in this enormous room.

Art is very much a part of our traveling; we still get excited when we see an exceptional piece and one that we might consider for our collection. Although we ran out of wall space several years ago, by moving things around and discovering new nooks and crannies, we have, up to now, managed to squeeze new works in. I enjoy nothing more than going to auctions, Basel Switzerland, Armory shows or Art Basel Miami Beach with art loving friends—the Feinsteins, Eisenbergs, Bernsteins, and Wolpovs—and later relate our art stories and discuss news of the art world over dinner. It's hard to believe, but I think I feel even more stimulated and excited talking about art than I do about sports. I'm sure the passion for art that Judy and I share will grow even stronger now that I have more time.

In August 2000, Bev and Bob Yaffe chartered a private yacht, with the purpose of cruising the Amalfi coast of Italy, with three other couples: Sue and Len Feinstein, Barrie and Bob Blumenthal and ourselves. Bob had set up an exciting itinerary. We flew to Sardinia and stayed at Calle Di Volpe.

The 60-foot luxury yacht, manned with an experienced crew, was named "Big D" and owned by the writer of the A Team, a successful television series, and other adventure movies. It had five bedrooms, a beautiful living room, dining room and a rear outdoor deck where we would eat lunches and hang. Amalfi, Sorrento and Capri were among the ports that we visited. Besides going to some great restaurants, we were served dinner on the yacht about three times. It was a very special trip, with very special people and one that Judy and I will always cherish, as it was about six months prior to Bev's passing. I always felt that it was our farewell trip.

On my sixty-fifth birthday in October 2000, we took our adult family to the Ritz Hotel in Paris for a four day planned weekend. Everyone was able to go except Tracey, who was pregnant with the twins. We had set up reservations at the finest restaurants months in advance and even planned to visit some unique dining places. I had a van driving us all over and we went to many museums, did the normal sightseeing, and of course, shopping. We took a tour of Moet Chandon and Don Perignon caves in Remes, set up by my friend, Michael Goldstein. We ate lunch at a beautiful five-star restaurant, Boyer. The trip culminated with the seven of us having my birthday celebration at Alain Ducasse; it was one of the finest meals that any of us had ever eaten. We crammed a whirlwind tour of Paris into four days, but it definitely wasn't an abbreviated course.

Earlier that year in March, Judy and I had just finished dinner in a restaurant and decided, rather than go home, to take a walk in the Mizner Center in Boca. As we were walking, I passed this huge model of an elegant apartment house that faced both the ocean and the Intracoastal Waterway. When we went in, a saleswoman told us that there was an apartment in the Excelsior, with a 12,000-foot terrace with a swimming pool on it. We had no intention of moving, but the thought of looking at this apartment excited us both.

We were about to leave to go back to New York for the summer, so the saleswoman suggested we meet the owner the next day. We spent two hours with him looking at plans and getting very excited. Back in New York, I signed contracts on this extremely expensive Boca apartment. We decided we were going to do it in a Balinese/Oriental style, so we went to see Juan Montoya, the well-known architect. After sitting through a one-hour slide show and listening to him, it was obvious that all he wanted was a signed check with no number on it, and *he* would do the design. He had no interest in what we wanted. I left there knowing that I could never work with him.

Margaret Kittinger of Beyer, Blinder, Belle had worked on a major White Plain's office recycling job that won much acclaim for Reckson, and I knew she was very talented and someone I could work with. We spent the next several months on creating plans and buying furniture and fabrics. As the apartment was nearing completion, we went through it to see the view. Initially, we were excited, but disappointment set in as we realized the ceilings were only eight feet high. However, we continued to move ahead.

That Thanksgiving, after working for many months going to furniture and design showrooms, Judy and I decided to spend a quiet vacation in our Boca home in Mahogany Bend. As usual, we went to dinner with many of our friends, who couldn't understand why we were moving. One day, Arlene Karyo asked the magic question, "What are you going to do after you look at the ocean?" It started us thinking that all of our friends were located in Boca West. We had been the first ones there and had encouraged about twelve other couples to follow us. The Excelsior was at least twenty minutes away. Judy and I started to make a list of pros and cons and, by the end of Thanksgiving, we were both sure that it was a mistake to move, even though all of our plans were finished.

The builder was able to re-rent the apartment, even though he'd done a lot of work on it, and returned my entire $700,000 deposit. He was a true gentleman.

We called up Margaret and switched gears, now redoing our entire Boca West house the same way. Judy negotiated a walk-in closet and a larger bathroom out of it, which meant we needed to extend the house and also needed a variance to do it. We left the walls but virtually everything else—the floors, the roof and anything in between—was changed. All of our new Oriental antiques and furniture now replaced most of what we had. When we finished a year later, we felt that we had created a *Shangri-La*, as Bali had been to us. With Margaret's help in designing the cabinets, ceilings and wall treatments, it became very special.

That Fall of 2000 at a family Rosh Hashanah dinner, Glenn told us he was offered an interview for a very prestigious job, that of Director of Contemporary Culture at the Royal Festival Hall in London. Royal Festival Hall is larger than Lincoln Center and the Brooklyn Academy of Music rolled together.

It would be a three-year contract. I knew it would be especially difficult for Judy not being able to see Milo, the twins, and Tracey. Even though it saddened us that they would be moving away, I felt it was important to Glenn both professionally and personally. I encouraged him to interview and do everything he could to take the job. Initially, he was turned down because they felt it was too big a position for an American to hold.

By Christmas, they had a change of heart and he was asked to come back and be interviewed by the board of directors because he was by far the most qualified candidate who'd applied. He flew over to London again and was hired. In the new year, Glenn and his family packed up and moved to London. He has done very well in that position and will probably stay for some time. Glenn and Tracey and the children visit us once or twice a year and we visit them in London once a year. We've also managed to go on a vacation together once or twice a year, but we still miss them an awful lot.

By 2001, the first of our very dear friends passed away—Beverly Yaffe. Bev had been fighting cancer for thirteen years. Bev and Bob had been our traveling companions, neighbors and close friends. We'd raised three boys together.

It was inspirational to see the truly heroic way Bev faced up to her illness for thirteen years, rarely complaining, always smiling and never taking a minute off from celebrating life. When she finally passed, it came as a shock to most of us because she had won so many battles that we never expected her to go. We had been out with her and many couples three nights before and I saw her as she passed our house on the way to play golf the day before. Memories of her are what brings a smile to my face, whether it was her infectious laugh when she entered a room, the close friendship with Judy, the way she *kvelled* when speaking about her grandchildren, Hillary, April and Levitt, or her insistence that someone in a movie or at a foreign airport shouldn't smoke in a non-smoking area. If they ignored her, she would remind them that she was there by smacking them with her pocketbook. That was Bev, and Judy and I loved her.

Two years later, the scenario seemed to repeat itself with Sandy Berlin, Judy's sister. Sandy and Norman had been close friends of ours since their marriage. This time the shock was that Sandy, who hadn't been sick, was diagnosed with a rare cancer and died within a few weeks.

Sandy had an amazing energy and vitality for life. She was a Top A tennis player. Having played her whole life, she won many club championships. Sandy loved the outdoors, loved to travel, and was the ultimate enthusiast over everything she saw or participated in. She loved nothing better than attending the ballet with Judy and Roz Goldman, a good game of doubles, or walking with Norman. Our memories of her will always be of smiling, enjoying, and talking about her grandchildren, Jackie, Shep, and Charles. I remember how sure Sandy would be on a subject, at times intimidating, but if you corrected her, she would just think it was funny that she was wrong and laugh. Sandy, as Bev, has a special place in our hearts. They were both special people who were taken from us and from their families far too early in life. This made us realize how precious every moment in life is.

In 2001, having had my green Bentley for three years, I was driving home from the city at about 9:30 at night. I was on Wolver Hollow Road approximately half a mile from home, listening to sports on WFAN. The next thing I knew I had either blacked out or dozed off. I hit a mailbox with the 6,000 pound Bentley. The airbags popped and the tire blew. About 200 feet past the mailbox I came to a wobbly, rickety halt. I knew it was a pretty violent collision, but I seemed okay. My knee was a little banged up and my arm burned from the airbag. The windows were smashed. Thinking I must have dented the bumper or maybe the mailbox hit the windshield, I got out of the car to look.

To my amazement, my car was in ruin. It was totaled. I was fine. It was only my second accident, but both were beauties.

I walked back to see what kind of mailbox had greeted me so abruptly, only to find out, with my luck, that it was right next to the road and on a six-inch steel lolly column. It was anchored into a 3' x 2' chunk of concrete, which the car had pulled out of the ground. As I was standing there, a police car unexpectedly pulled up. He turned on his red flashing light so no one else would hit me. The car phone still worked so I called home because I knew Baldo, our house manager, would be there. As I stood on the roadside with the police officer, cars slowly going by, I saw Judy drive past. I tried to wave her down, but she was oblivious to the whole situation. Baldo came to get me.

I needed a car right away and felt that I was no longer happy with the quality of Mercedes. I went into the city for another Bentley. They told me that they had a unique model and color coming in, one that Bentley had made only for the Queen. I was heartbroken that I had ruined my green Bentley, but now I would know what it would feel like being a queen.

*"I have the simplest tastes. I am always satisfied
with the best."*   —Oscar Wilde.

In recent years, we have become close with a group of friends that we met originally through the Yaffes. We all lived on Long Island and now we have vacation homes in Florida. One New Year's Eve about 8:00 P.M., Judy and I, Iris and Stanley, and Bob Yaffe and Lois Bennett (who had recently started seeing each other) were on our way in a limousine to Frenchman's Creek, about an hour's ride from Boca. We were going to Bonnie and Larry Fenster's annual New Year's Eve party, where we would meet the rest of our friends, Sue and Len Feinstein, Barrie and Bob Blumenthal, Joan and Jerry Scheckman and Ilene and Stan Barshay, who all live in Frenchman's.

We were going along at a nice clip on P.G.A. Boulevard in a stretch limo, almost there, when all of a sudden I saw a car jumping the divider and heading full speed, out of control and straight towards us. It had to be only 75 feet away. I was facing the front seat, and Lois, who was sitting next to me, also saw it and stiffened. As usual, I remained relaxed, like during the near plane crash. Our driver reacted immediately and yanked the limo across the boulevard. Something hit the rear of our car as the other car passed and then crashed into a tree behind us. Everyone was jostled, and we all agreed it was a close call and we were lucky to be alive.

The party, as always, was special and this close call made us aware that life is even more special. We skipped the following year's party to avoid the drive, but this past year we couldn't stay away.

In the summer of 2003, after almost four years without all the boys and their families all joining us on the same vacation, we were finally able to get all three families and Judy and myself together. We took everyone to Capri and had a wonderful time. The cousins had a ball together as they always have. We had a going away dinner on our last night, and at the end of the meal Milo started to cry, upset that he had to say goodbye to his cousins. Milo hugged Benjie, and Benjie started to cry too, then everyone broke out in tears.

We had stayed in Capri for eight days, and then everyone else returned to Long Island, except Judy and me and Glenn and his family. We went to Sardinia to have some alone time with the twins and Milo who had, up until then, been distracted by their cousins. Three days with Tracey and Glenn at Casa de Volpe in Sardinia was a lot of fun, and we really bonded with Milo and the twins.

It's not easy getting everyone together, but it sure is wonderful to have our children and our grandchildren all together on a vacation with us. Hopefully, we will do it many more times.

To celebrate our forty-fifth anniversary in 2003, Judy and I, along with our friends, Iris and Stanley and Peggy and Bruce, decided to take a West Coast trip. For the past five years, I have had a fractional ownership in both the Hawker 800 (eight passenger) and the larger Challenger, which is a nine-passenger jet with a lot of luggage room. We flew across country in the Challenger to Las Vegas, our first stop, where we stayed at the Bellagio for three days, caught all of the top shows, including Dion, and ate in the new hot restaurants.

We then flew to Napa for three days, where we stayed at the Auberge du Soleil, and traveled around the area in a stretch chauffeured SUV with room for twelve. As a group we visited many wineries, getting a V.I.P. tour at Opus through our friend, Michael Goldstein, from Park Avenue Liquors. I went to even more wineries and replenished my wine cellar, buying several cases of boutique and hard to get wines. Everyone bought some wines using me as their "Maven." I have collected wines for years, tasting most but rarely drinking a full glass because of my allergies. We dined at a very special restaurant, Tra Vigne. As I walked past a table, I saw a young woman sitting alone, moaning as if she was having an orgasm, and I had to stop and ask what she was eating. She said, "Short ribs" with a smile . . . . I knew I had to have some of that! The short ribs were outstanding, as the chef cooked them two different ways for twenty-six hours. It was one of the best dishes I have ever eaten.

The highlight of our trip was dinner on our last night in Napa at the French Laundry. Our travel agent Judy Stein, after much persistence, managed to secure a hard-to-get reservation. That was our forty-fifth anniversary celebration dinner. We knew it would be one of the most expensive restaurants we had ever been to, but none of us were disappointed. It lived up to our expectations and we all had a wonderful meal. After dinner, we were given a tour of the kitchen and an autographed cookbook from the chef, Thomas Keller.

We then left for San Francisco where each couple did their own thing. Judy and I went to museums and then walked through the art area, finding a glass gallery that carried all European artists. We bought several pieces there.

The last leg of our trip took us to Cabo San Lucas, in Baja, where we stayed at Las Ventanas. We all had great villas, although as usually was the case, Peggy sampled several . . . definitely a case of Feng Shui. It was a romantic, beautiful spot with an elegant clientèle. The lunch and dinners were all enjoyable. After five restful days, we returned on the Challenger to New York.

CHAPTER 21

Full Circle

After spending two years refocusing after the Front Line write-off, our business began to feel the impact of the recession and the aftershocks that resulted from the corporate greed that plagued the nation during the boom of the 90s. Two of our largest tenants, WorldCom and Arthur Andersen, were rocked by scandal and forced into bankruptcy and liquidation. We got back over 300,000 square feet of high quality space from them and our occupancies were hammered in what was already a tough market. As the real estate markets got tough, the investors began to question our strategy and operating structure.

The market was unforgiving and our stock suffered accordingly. Wall Street analysts felt that we had too much family in the business and that our industrial portfolio didn't offer enough future growth. Our overhead was high because of the way we structured our company with the belief that real estate is a local business. Each of our four markets had its own organization and managing director and its own offices. This worked well when we were in an expansive market, but the higher overhead weighed down the company in a stagnant market. We had built a very strong organization and now, in order to better deal with the economy slowing down, we began to sell off problematic and no growth properties.

By this time, the Sarbanes Oxley Act had been passed into law. After all the Wall Street scandals and the burst of the Internet bubble, this law now put all sorts of restrictions on how business could be done as a public company. Now, more then ever, I found myself working less and less on real estate, which was what I really loved to do, and more and more on Wall Street matters.

The only real estate activity that I got to spend much time on at this point was our annual reports. We won four prizes for our annual reports, one for top overall annual report in the entire market. I was starting to have serious doubts whether real estate and the stock market were compatible. Real estate is a cyclical, long-term business where financial flexibility is imperative. The stock market is short-term oriented, looking for quarter to quarter results in a business that develops over years. In addition, Reckson's strategy was limited by its investment grade ratings. I never believed that it made sense for a real estate company to have investment grade ratings in the private or public sector. I could see how going public changed an entrepreneurial business into an institutional one. Mitchell and Gregg told me they felt frustrated by the public markets and were ready to move on to other interests.

It was in this climate that I came to the decision that it was time to explore a merger or a sale. I felt strongly, as I did about most things once I made a decision, that we should be proactive. At this point, over 50% of Reckson Associates' shares were owned by five institutional funds.

Speiker, a West Coast REIT (comparable to us) had just sold to Equity Office, the largest office REIT in the market.

They were more of a holding company than a builder/operating company. During the sale, Speiker set a precedent by negotiating a tax structure which really made it possible now for the family to avoid catastrophic taxes upon a change of control. That was good news.

Many of the REITs had hit new highs, and one of these was making overtures to us about merging. Our stock remained low with the same problems that had plagued us: corporate governances—the biggest one being that there were too many Rechlers on the Board of Directors and in the management of the company. We always heard "family business" when things weren't going well. We were being approached by another REIT and an M & A banker (merger and acquisition banker).

After the Front Line problems we had changed our law firm to Wachtell Lipton, a top-of-the-line corporate firm. For all the reasons I outlined above, and also because interest rates were low, I felt the timing might be right to sell the company.

We put together an experienced team of professionals, which included our counsel from Wachtell Lipton and tax attorneys Goldman Sachs, to advise the family, and Citigroup to advise the Board of Directors. Peter Quick, who had been Chairman of the American stock exchange, had joined our board the year before and served as lead director. He and Lew Ranieri represented the independent board members and headed the committee that would establish the minimum price that Reckson was willing to take with management. They gave us the okay to proceed developing the necessary analysis to evaluate alternatives.

As expected, we were approached again about a merger. This time, the whole board sat down and talked about why we thought the timing was right and how even the talk of a merger might unleash value in our stock. Connie Stevenson, a Board Member since our IPO, knew the personality of the bidder, and told us they had a reputation of waiting until the last minute and then re-trading at a lower price. Lew Ranieri agreed and didn't think these people were even capable of closing a deal. I was wary, but felt it was important enough to start negotiations and see where they led.

True to form, it was a fight and naturally the buyer wanted to pay a lower price, but it was just another day in a typical business negotiation. We prepared total numbers on each and every building and presented it to the bidder without any backup, just for their interest. We told them if they hit our price and met our terms, we would then proceed with the deal and give them the backup. We finally settled on a price close to our original, and then proceeded to give them all the backup information that supported it.

The deal still was complicated by the industrial portfolio and RSVP (Reckson Strategic Venture Partners). The other company wasn't interested in any of the industrial properties, and wouldn't agree *not* to sell them later. As our operating partnership units were tied up in the industrial portfolio, we couldn't sell them because it would trigger significant taxes to the family. We had to put in a clause that would prevent them from selling. Speiker had done that, so we knew it was possible. The other problem was RSVP since the bidder was concerned about the value of its assets. In order to make the deal simpler for the purchaser, the family agreed to guarantee a minimum price if they were unable to sell RSVP at a higher price.

Negotiations dragged on much longer than they should have extending over six weeks. Our lawyers and bankers were amazed at how little real estate sophistication the other company's management team seemed to have during the negotiation. This made us question whether these people should be allowed to take over management of our company, whether this merger would be good for Reckson, for our shareholders and for our family.

The night before the contract was to be signed; Scott got a call low-balling the price. He told them to send all our papers back and refused to negotiate any further, thus ending the deal. Connie and Lou had been right. These people were only buyers of distressed real estate.

Sometimes the best deals of all are the deals you don't make. However, some good did come out of our arduous experience with the merger process. The work had now been done and a road map created. In addition, the board received a credible third party offer on the Long Island industrial portfolios, which would prove to be extremely important as we turned to Plan B. The family and management held a few strategy sessions, but it was obvious that we still faced the same challenges. The merger talks hadn't changed our situation, and more talks with other investors were not likely to yield better results unless we got rid of the stumbling blocks.

The main problem that still confronted us was the family's operating partnership units were tied to the Long Island industrial office portfolios and any sale of these separate properties would trigger horrific tax consequences. We couldn't change the way the market valued us. According to them, our corporate governances were not investor friendly—we had too much family on the board and in management; and we offered no flexibility to sell the industrial portfolio because of the tax situation; and no ability to trade *top:* Fourth of July 1995, top row L-R: David Blumenfeld, Rabbi Marc Gellman; center L-R: Anna Blumenfeld, Mitchell, Debbie, Kathie and Alan Yaffe; bottom row Lisa Rechler, Jackie, Mark, Gregg Rechler.

bottom right: Annual ice cream man with Norman Rothstein, July 4th Terrapin. Jade (2) al fresco

bottom left: July 4, 1996, L-R: Willi, Benji, Judy holding Jade, and me.

RECHLER REVELRY
It Gets Better Every Year!

top: Father's Day 1996, top L-R: Willi, Mitchell and Ben, me and Glenn; front L-R: Gregg and Wiley, Scott and Gabrielle, Roger and Bill, Todd, Mark and Jade.

bottom: Terrapin Hill, the lower pool on the last Fourth of July party, 2000.

top: The Over-the-Hill Gang, the Dix-Hillians: L-r: Bob and Madi Kaplan, Judy and me, Bob and Bev Yaffe, Helene and George Barasch, around 1994, Jackson Hole, WY.

bottom: Fortieth Anniversary celebration on the Silver Seas, a three week trip from Australia to Bali. L-R: the Rechlers, the Tuckers and the Rabinowitzs, 1997.

top: First trip to St. Barts, 1997, at Lafayette—needed a mortgage to eat there!

center: Mark, Glenn and Mitch at St. Barts.

bottom: St. Barts, L-R: Jade, Benji, Willi and Milo and me.

top: Sixty-fifth birthday celebration in Paris with our family in the caves at Reims in 2000.

center: Me with Tahiti Momma - Millenium cruise around Tahiti with five couples, the Tuckers, the Rabinowitzs, Rita Ullian and Chuck Karst, Arthur and Thelma Muskin, in 2000

bottom: Always rode on western vacations. Together with Benji in the Tetons, Wyoming in 2001.

top: Family trip to Anguilla, February 2001. Glenn and his family had just moved to London and couldn't join us so Iris and Stan came along.

center: At the Phoenician in Scottsdale, Arizona, L-R: Mark, Glenn, Mitchell and I with the girls.

bottom: All the family was on the trip to Capri, August 2003.

top: Mitchell and Debbie

center: Mark and Jackie

bottom: Ruby, Glenn, Sasha, Tracey and Milo, living in London (visiting Paris).

JoAnne

left: Willi at Terrapin
Hill at eight.

right: Benji at
Terrapin Hill at six.

bottom: Ruby and
Sasha at age three
and a half.

top left Milo, English schoolboy, 2002.

top right: Jade at seven.

bottom: Capri 2003, Willi, Sasha, Milo, Ben, Dylan and Jade

top: Me with Jade and Willi at Capri.

center: Willi, me, Ben and Barney at Terrapin.

bottom: Willi at four with me at Terrapin.

top: The ladies; Sasha, Jade and Ruby enjoying gelato in Capri.

center: Milo and Ben.

bottom right: Capri 2003, L-R: Sasha, Ruby, Willi, Jade, Ben, Dylan and Milo.

bottom left: Dylan and Henry

top left: Ben and Dylan at the 45th anniversary celebrations.

top right: Jade and me at the 45th anniversary celebrations.

bottom: Sasha and Ruby with me, Barbados, 2003.

top: L-R Len Feinstein, Jerry Scheckman, Bob Blumenthal and me

center: standing L-R Bruce, Judy, me, Iris and Stan, seated, R-L Cindy and Norman Rothstein

bottom: Mitch's family, Judy and me and Mark's family on our 45th anniversary.

top: Willi's bat mitzvah, L-R: Bob and Lois Yaffe, Susan and Leonard Feinstein.

bottom: Willi and her brother Ben with cousins: L-R: (front) Dylan, Sasha and Ruby, (second row) Ben, Milo and Jade.

top: Willi's bat mitzvah on March 28, 2003. We are proud grandparents.

bottom: Judy and me with our three sons at Willi's bat mitzvah.

top: Judy and me with the kinder L-R Willi, Dylan, Jade, Sasha, Ruby, Ben, and Milo.

bottom: Mitchell and Debbie, Glenn and Tracey and Mark and Jackie.

up to class A real estate, preferable in the New York City market, which was a favorite with investors.

Within a couple of weeks, we came up with a responsible solution to our problems; one we thought would be best for the company, its shareholders, and ultimately our family. The family would acquire the industrial portfolio from Reckson, and three Rechlers would exit the company. That way, Reckson would be able to pursue its strategy and be more attractive to investors. At the same time the family would not have to worry about adverse tax consequences.

We talked to the independent board members and many were upset to learn we were considering leaving because there were too many Rechlers. John Klein, who had known our family since my father's days said, "You can't have too many Rechlers for me. It's the only reason I'm here." Feeling good at that, I found myself tearing up as others voiced their agreement. However, in the final analysis, having Rechlers leave Reckson was what was best for both the company's stock price and the family.

We realized that the family's acquisition of the industrial portfolio from the company was going to come under significant scrutiny and attract lawsuits, so everything had to be at arm's length and follow a well thought out process. To make this deal work, the independent board put together their own professionals. They had Citigroup prepare a fairness opinion, Cushman & Wakefield prepared separate appraisals, and Wachtell acted as their attorney.

The family hired Fried Frank and Blank Rome as our lawyers. Because management was involved every "I" had to be dotted and every "T" crossed, and our offer had to be higher than any others we had received. We made an offer to acquire the industrial portfolio with partnership units and cash; but the independent board led by Peter Quick felt the price should be based strictly on the appraisals for the deal to be at arm's length and in good faith. In order to finalize the papers we dropped the land, we were in the process of purchasing out of the deal, so we paid more and got less for our money than we had originally hoped. But, we were certain our major shareholders would welcome the solution we had come up with.

Only sixty days after the original merger talks with the other REIT had broken up, we signed the deal, trading our business units and cash back for the industrial portfolio. The family members protected their business units by trading them back into the industrial portfolio and going private. In doing so, we had given up a good portion of the contracts we were entitled to at a large savings to the company. As expected, there was an initial firestorm of bad press. A frivolous lawsuit was put together by less than 1% of the shareholders accusing us of not paying enough because management bought it. We were confident that we'd made a good, fair deal. All the major investors thought it was a good solution under the circumstances.

We changed our corporate governances to allow Reckson better alignment with investors. We changed the Board of Directors to six independent directors and only two inside directors instead of five—one being Scott and the other myself, as non-executive Chairman. In order to vote the entire slate of directors each year, we got rid of the staggered board. We modified the ownership limit so it could no longer be used as an anti-takeover tool, and made other changes to corporate governances that were also improvements.

As the first of our peers to address the misalignment in REITs, we eliminated all operating partnership conflicts. Other REITs were in the same situation—the principals might hold certain real estate in partnerships and then years later the REIT would decide they wanted to sell that original real estate, but selling it would cause substantial tax damage to the principals. All in all, Reckson had one of the best corporate governance structures in the REIT universe.

The public company was able to reduce a total of $7.5 million in overhead. Some of it came from Rechler related management salaries and the rest was industrial overhead that went with us. The company now had the flexibility of selling any of their office buildings without being tied down to the operating partnership units. By the time of our closing a few months later, a large majority of the market had praise for the deal.

Scott was the sole C.E.O. and President of Reckson. Todd Rechler remained Director of New Jersey and I continued as Chairman in a non-executive position, no longer active in management. The family still owned a significant amount of shares in the company. The New York City real estate market was hot, the company was able to promote management and the shareholders were happy.

The family closed with Reckson paying $315,500,000 for ninety-five industrial buildings on Long Island. On the same day, the REIT used that revenue to buy a major Class A, 46-story, 1.1 million-square-foot office building for $321 million: 1185 Avenue of the Americas. Most of the shareholders were ecstatic, because shortly after that, the stock began to rise. Three months after the closing the stock reached an all time high of $28 per share. The company made a lot of money, issuing stock—not at what the family had valued the industrial portfolio for at $23—but for 20% more. The family and three industrial REIT partners no longer had devastating taxes hanging over their heads.

In January 2004, the family formed a new business called Rechler Equity Partners, a classic win-win for all partners. The Rechler family had come full circle and was back in the Long Island industrial business, which my father had started in 1958. We now were the second largest real estate company on Long Island. Of course, the first was Reckson, but it was now a purely regional office REIT.

"Good fortune is not the absence of problems, but the ability to deal with them."   —*Donaldism.*

CHAPTER 22

The Last Word

I've been known to take strong stances and be opinionated. As I said, I call it the way I see it. This book wouldn't be complete unless I commented on how I see things because I'm usually right. *That's just how it is.* (All my friends will get that, and the fact that since this is *my* book, I'm having the last word.)

Integrity and Ethics—Having been in business for 50 years, I found the period from the mid-eighties to the present changing the way people did business, as well as affecting other walks of life. Integrity and ethics have been declining on a steady basis. The nineties saw it crash down with a mighty crescendo. This decline hit all walks of business, but the greed of the public sector, by its very size and nature, far exceeded most business areas. A close second occurred in sports, particularly with the athletes. The decline in ethics is evident in both business and sports: by the athlete's disregard for the fan or the executive's disregard for the investor, and by cheating and disrespecting the sport or institution they work for. I feel there was a breakdown early on, fueled by greed and ignited by the thought that the world owes them something.

It starts with our higher institutions of learning, where thousands of subjects are taught the credo: "succeed at any price," but rarely is a course in business ethics on any curriculum. It's unfortunate that the government has to set up all sorts of watchdog rules in order to deal with these problems. Fortunately, there are still people, firms and players who don't do what the others do just to have the same advantage. They have the integrity to do what's right.

My Politics—In 1956, when I was a graduate from college I voted in my first presidential election. There were fifty-two lobbyists then in Washington. In 2004, as I am about to vote in my twelfth presidential election, there are over 25,000 registered lobbyists in Washington. This, in part, has become a big problem with our political system.

Looking back on the first five presidential elections I voted Democrat for my first twenty years. When you are young, you are a liberal. It is said, if you're not a liberal when you are young, you have no heart, and if you are not a conservative when you are old, you have no brains.

At the same time I was voting Democrat in presidential elections I was voting for Rockefeller and Javits, both Republicans, as governor and senator respectively of New York. I consider myself an independent, a centrist. Like the majority of Americans, I don't support anyone too far to the right or too far to the left. Over the next twenty-eight years and seven presidential elections I pretty much split my vote, not wanting one party or the other to control the country too long.

"In America anybody can become president. That's just one of the risks you take."—Adlai E. Stevenson

War—Every war brings death, destruction, degradation and cruelty. To consider one good, worthwhile and heroic just goes against its very nature.

So World War II was a model of a good war. It was easy to differentiate between good and evil and black hats and white. We freed Europe from tyranny and helped them recover, post war. The G.I.s were totally committed. The American people were an intimate part of it, working in defense plants. Women gave up nylons, saved balls of tin foil and cans of grease. Men gave up gasoline and served as air raid wardens. Families had meat rations, scarcity of foods and drove in the dark. Children collected tin cans and bought Liberty stamps at school.

On the sixtieth anniversary of D-Day, I'm writing this while watching the dedication of the new memorial to the 400,000 G.I.s who died in World War II. Several years ago I visited Normandy with Judy, and I am in awe of what they sacrificed. It renewed my memories of the radio news and Life magazine photos and stories I heard and saw as a boy growing up during World War II. I will be forever appreciative and humbly grateful to those G.I.s.

Today, more than a year into the war in Iraq, and three years after 9/11, the U.S. is divided at home and ridiculed abroad. The American people are, for the most part, detached from the war. The clarity that was available in WWII is not visible today. The intelligence and information seems to have been flawed, though the goal today remains clear and the fortitude of the troops in Iraq is still inspiring.

While America has a vision of a new Iraq arising out of darkness, we're not sure whether it will end in victory. America, never the occupier, may want to give the gift of freedom to the Iraqi people, but it will be a failure unless it's taken, won or earned. Our post war planning and management has been poor. It was our hope to set up a free democracy in a part of the world that had none.

As in WWII, the cause was just, the foe was a tyrant and we were attacked, but what remains to be seen now is if the reason is believed to be true. If not, I can't help but think of what Rudyard Kipling said of wars past, *"If any question why we died, tell them, because our fathers lied."*

Friends—Good friends should never stand on ceremony. Many of the same ingredients that go into a good marriage go into a good friendship, such as support in times of trouble, celebrations in times of joy, and humor—being able to laugh with and at each other. You must be able to communicate and be a good listener, and most important, to be able to go to friends in times of trouble and accept advise, compassion and sympathy in good faith.

Good friends are precious commodities and should be treated as such.

A perfect example of celebration, humor, and being able to laugh with and at each other was the Roast Toast on my sixtieth birthday trip by my good friends.

Donald's 60th Birthday Toast Roast

10 years ago the Riveranda took us for a sail
Now for a 60th, Donald is taking us to Scottsdale!

He's trying to make cowboys of 33 Jews
This battle he's going to loose.

On our horses we will go far
We're only used to riding in foreign cars.

Shopping is one of Donald's great thrills
The Scottsdale merchants love all those bills.

They're going to have their greatest year
Now that "The Donald" is here!!

On Monday night we know where Donald will
be Sitting in front of that big screen TV.

Come hell or high water he's going to watch his
          team play
Nothing else can get in the way.

Judy can dress as the most alluring dame
It's no use when there's a Giant's football game!

Of course the boys should have no doubts
They were conceived during the time outs!

Judy has put up with all Donald's shtick
And most of the time that's not an easy trick.

A toast to Donald will close this roast
But to Judy a medal is deserved the most.

So for this wonderful weekend we want to
          thank our host
By raising our glasses and making a toast.

With Love,
Madi & Bob, Arlene & Richard, Bev & Bob,
     Susan & Eddie, Helene & George

Health—We tend not to pay much attention to our health when we are young. I was just over forty when I started to realize my warranties were running out. As we get older, we spend more time visiting doctors, taking tests, and even talking about health, sometimes to a fault. As in most things in life, I try to be proactive and responsible for my own health, know my own body, and know what makes me feel good or bad, such as environmental things, exercise, diet, medication, and vitamins.

Because something works for someone else doesn't mean it will work for you. Listen to what doctors tell you, then investigate further. Make sure you make your own decisions. Be aware and check medication. If you have to go for surgery, make sure you talk to the anesthesiologist yourself. If you are in the hospital, check and make sure you are given the right medication. And if possible, it is a good idea to have or appoint an advocate for you in the hospital.

Family—As your children progress through childhood and become young adults, at some point you have to step back and realize that you've already given them as much love and advice as possible. You've taught them right from wrong and hopefully set a good example. Now comes the hard part. You have to back off and let them find their own way, while remembering they are not you.

I am proud of my sons and the men they have become. They also did a good job picking daughters for us. No matter what the age of your child, as a parent you should always be there to offer love, support, and advice when asked for, and on rare occasions, even when not asked for. The only thing any parent can expect in return is love, respect and visits with the grandchildren.

We enjoy and cherish our grandchildren. When the twins, Sasha and Ruby, are smiling it is like the sun is shining bright, and when we hear their little British accents it melts us. To watch Jade perform, whether on stage or in the kitchen, is a joy. She is a little diva. Ben was born with a jock on; there is no sport that he doesn't take to naturally. It is a thrill to watch him play because he is the ultimate ballplayer. It is also a lot of fun to spend time with him and you can't help but get caught up in his infectious laugh.

Milo excels in school and is a bright young man. It amazed me how patient and brave he was at handling an accident where he cut himself and we had to take him to the hospital for stitches. He plays soccer in London and is glued to the TV whenever a soccer match is on. On a recent visit, Milo watched every ball game with me and we went to a baseball game together.

I can see as I write this that five-year-old Dylan will be quite the ladies' man. He's just coming out of his shell and has become a boy who loves soccer and karate. Isn't that a pisser and somewhat sweet revenge—my two musician sons having sports enthusiasts for their sons, definitely a generation skipping gene.

Willi, always a beautiful young lady, is accomplished in piano and tennis. When she performed at her Bat Mitzvah last year, she was so outstanding it sent a chill down my spine and brought tears to my eyes. It was easy for all to see how she radiates from her inner soul, projecting consideration, compassion, and kindness.

It is said that grandchildren are God's reward for surviving *your* children.

Organized Childhood—A subject that amazes me is organized childhood. Today, children are so busy with organized team sports—sometimes three or four different sports—that all parents do is carpool them around. When they're not playing team sports, they're taking tennis, golf, or karate lessons, and dance or music lessons. In any spare time they have, they're all going to each other's birthday parties. Not to mention, there are Hebrew lessons and other religious schooling. Any time left over they probably spend on a computer, or with Play Station, which also gives them instructions.

Our generation chose up sides, created our own games, settled our own arguments and used our imaginations. We had the freedom to succeed or fail and learned to deal with it. What

I really wonder is if the spontaneity, communication skills, creativity and innovativeness of previous generations could be lost on the children of the Twenty-First Century. It will be interesting to see what the future holds.

Charities—Since the age of thirty-five, when I started being active in real estate, I became active in several charities. I always felt it was important to give back to the community, not just the one where you earn your living, but also where you live. Sometimes that is one and the same. If I recall correctly, the first board I sat on was the Long Island American Cancer Society. I've also been on the boards of North Shore Hospital, Tillis Center of Performing Arts, Friends of The Arts, Long Island Philharmonic, Nassau Museum of Art, and various others. Over the years, I was honored by a number of the charities. As it became tough to find time, I cut back on my board involvement, but I still support the organizations. I am currently very involved in the Nassau County Museum of Fine Arts as we are finalizing our plans for a major expansion.

As each of the five boys came into the business I encouraged them to get involved with a charity of their choice, whatever they liked. They all got involved—with not one but a number of them—actively participating on boards and contributing both time and resources. I'm as proud of my sons and nephews for their charity work as I am for their business aplomb.

Mitchell was honored and on the board of the Long Island Children's Hospital and president of the board of Friends of The Arts. Scott is involved with The Children's Museum; he was president and instrumental in getting the building built. He is involved with other charities, including some in Manhattan. Mark was involved with Long Island

Harvest and worked in soup kitchens occasionally. Gregg's charity involvement is with the Long Island Breast Cancer Society and Long Island Air & Space Museum, as well as the Nature Preserve. We were instrumental in opening the new buildings of The Children's Museum, the Space Museum and now, hopefully, Nassau Museum of Fine Arts.

Money—Money is not a form of keeping score. It doesn't purchase respectability. It doesn't make true friends. It doesn't endow you with class or guarantee integrity, intelligence or accomplishments. Money is a tool that if used properly *helps* you to obtain some of those things, as well as comfort, education and security for your family. It is also possible to use money to gain satisfaction through helping others.

"Not everything that counts can be counted and not everything that can be counted counts."

—*Albert Einstein.*

Money doesn't buy happiness, but like chicken soup, it doesn't hurt either.

Sports—After World War II, New York City was the center of the Universe. It was where everyone—writer, dancer, and artist —would come to succeed. The city was affordable, and besides being the financial capital of the world, it became the cultural capital of the Universe. In the center of all this were three great New York baseball teams that gave the period from 1947 to 1957 a name; the era was known as baseball's "Golden Decade."

It started with the arrival of Jackie Robinson, which was an experience of enormous intensity and significance. It had such other notables as Leo Durocher, Casey Stengel and Jolting Joe DiMaggio. There were three other Hall of Fame center-fielders: Willie, Mickey and the Duke (Mays, Mantle and

Snider). Catchers were Yogi Berra, Roy Campanella and Wes Westrom. The Hall of Fame shortstops were Pewee Reese, Phil Rizzuto and Alvin Dark, and pitchers such as Preacher Row, Carl Erskine, Allie Reynolds, Don Newcomb and Whitey Ford. A bevy of Hall of Famers played in New York.

A New York team was in the World Series every year, and *seven* times in this "Golden Decade," two New York teams played each other. To boot, the three great broadcasters expounded the glory of radio: Mel Allen (Yankees), Red Barber (Dodgers), and Russ Hodges (Giants). Baseball talk has never again been as passionate.

The era was magnificent and it had a certain symmetry. It began with Jackie Robinson arriving and ended with Walter O'Malley departing with the Brooklyn Dodgers—the hero and the villain who hated each other. If you were a baseball fan, you had to be there to feel it and understand it. There never was a time or place like it, and probably never will be again.

By the time I was five years old, I really started to know players and follow them. I still remember Dodger games I saw at Ebbet's Field in 1940. All my life I enjoyed watching sports, although I was definitely a "homer," and it was rare that I would watch another city's team play.

There is nothing like being there to witness the great moments in sports. I've been lucky enough to be at such classic sporting events as seeing Jackie Robinson steal home to win a World Series game; Don Larson pitch a perfect World Series game for the Yankees against the Brooklyn Dodgers in 1956; and I was at the Garden in 1973 when Willis Reed limped onto the court for Game 7 against Wilt Chamberlain's Lakers, creating such an emotional high that even though he played about five minutes, it propelled the Knicks behind Walt Frazier to the Championship.

I saw two New York Giants' championship football games in the mid-50s when the Giants were led by Frank Gifford: One at the Polo grounds and the other at Yankee Stadium. Debbie, Mitchell and I traveled to L.A. and Tampa twice for three Superbowl games. During 1987 we saw Phil Simms break the Superbowl passing record while Lawrence Taylor, one of the greats of all times, terrorized the opposing quarterbacks. During the first week of Desert Storm in the 1990 season, after waiting three hours in a security line, we watched the Giants win their second Superbowl, when with seconds left, Norwood, the kicker for Buffalo, on a short 32 yard field goal, was wide left.

In 1980, we saw Bobby Nystrom score the overtime goal that propelled the Islanders to the first of their four Stanley Cups. In the sixth game of the 1986 World Series, Roger, Mitchell and I were at Shea Stadium, and watched with jaw-dropping amazement as the final out to end the Met's hope bounced through Bill Buckner's legs, only to see the Mets win the series.

You never know how a game is going to end. In that way, it resembles life. I could never understand how you can prefer a symphony, concert, movie or play when you know how it ends before it begins. How in the world can anything compete with the spontaneity, improvisation and the sheer excitement of an athletic event?

Heroes and Legends—The first half of the Twentieth Century probably had as many legends as any other era in history. By the time I was finishing high school, a number of these strong individuals had become my heroes and would remain that way for the rest of my life. They were all problem solvers and had the courage of their convictions.

Albert Einstein graduated a mathematician. He was refused a teaching post in Switzerland because he was Jewish. Einstein produced much of his remarkable work on the Theory of Relativity in the evening, during the time he was working in a Swiss patent office during the day. In 1914, he was appointed professor at the University of Berlin, where he wrote his monumental paper on the Theory of Relativity. Einstein renounced his German citizenship in 1933, as the Nazis were on the rise, and became an American citizen. He was an accomplished violinist who loved music. He was also one of the greatest scientists since Newton, who wrote about philosophy, politics and war. His name was synonymous with genius. Some of his philosophical quotes are: "Try not to become a man of success but rather a man of value," and "The only source of knowledge is experience." He believed that intellect had little to do on the road to discovery. He explained that there came a leap in consciousness, call it intuition; the solution came to you and you didn't know how or why.

Even as a young boy, when I listened to Winston Churchill's oratory I knew there was something special about it. He was eloquent, inspiring and heroic. In the 30s, he was the one voice that stood in the wilderness, trying to awaken Britain and the world against the threat of Hitler and Nazi Germany. He was looked at in the 30s, prior to the Nazi aggressions as an eccentric hawk, but in truth, history proved him to be one of the great visionaries of the times. He was a man of many skills, never without a cigar, painted over 1,500 oils and wrote several volumes of history. He had a sharp wit and was the master of insult, of which Clement Atlee, head of the opposition Labour

Party, was his favorite target. For example, he would counter Atlee with: "A modest man, who has much to be modest about," and my favorite, "An empty taxi arrived at 10 Downing Street and when the door was opened, Atlee got out." When the British people voted Churchill out right after the war, he said, "History will be kind to me for I intend to write it." And it was, no matter who wrote it.

Another hero to me was Pablo Picasso, the greatest artist of not only my time but in my opinion, of all time. He was never satisfied with his successes and would always move on to try something new, leading the way in what he did. He influenced me in my collecting and love of art.

As I look back at these men that I admired and who have influenced my life, they all had one thing in common: they weren't afraid to march to their own drummer, even though it wasn't popular at the time.

"Courage is grace under pressure."—*Ernest Hemingway.*

Love & Marriage—Marriage I liken to a great recipe. If you leave out too many ingredients or don't use the proper quantity, it is usually a failure. The main ingredient, of course, is love. But patience, support, affection, commitment, humor, communication and the ability to listen are all necessary. (I know what the women reading this are thinking: you have to *hear* as well as listen.) A most important ingredient is friendship. It definitely helps because fifty years is a long time, a lot of togetherness, and if you're not friends . . . .

I had the good fortune to find my soul mate, my love, my wife, at a very young age. Our recipe has endured the test of time.

While writing this, I did an impromptu survey using Judy's guest list from a party. Of twenty five couples who are our closest friends, most of whom we've known for over twenty years and a few close to fifty years, an amazing twenty four out of twenty five couples are within several years of celebrating their fiftieth anniversary. Only one set of our friends has been divorced. As I looked past this first circle of friends at our next group of friends, the results were the same. I guess our generation did something right.

Speaking of recipes, two other recipes that have survived the test of time are those we've mentioned: Judy's famous family peach cake and my sangria that we've made on the Fourth of July.

Life—Needless to say, I love the era and the generation I have been a part of. It has been historical, innovative,

---

DON'S "LA FITE'S" SANGRIA

Recipe: 8 quarts

4   bottles of red wine (I prefer Chilean La Fite's Rothchild)
Juice of 6 lemons and / or limes
6 oz apricot brandy
6 oz casis or berry brandy
6 oz Grand Marnier
10   peaches
4   apples
5   oranges
4   pears
* Allow fruit to soak in wine and brandy 24 hours before serving if possible.
* Add before serving:
10 cups   fresh orange juice and half peach juice
Note: halve the recipe for 4 quarts. Double recipe for 18 quart drums.

JUDY'S FAMOUS PEACH CAKE

Dough

3 cups	flour
1 1/2 Tbl	baking powder
1 tsp	salt
3 Tbl	sugar
1 1/2 sticks	butter, softened (or 3/4 cup)
3	eggs
3/4 cup	milk

1 17" x 12" x 3" Oval pan

Filling

Approximately 14 large ripe peaches.

Topping

2 1/2 cups	sugar
2 Tbl	cinnamon
1/2 stick	butter (or 1/4 cup)
1 pint	heavy cream
3	eggs (2 whole, 1 yolk)

Directions
* Preheat oven to 400 degrees.
* Grease a 17" x 12" x 3" oval pan.
* For Dough: In food processor, mix flour and butter to a consistency of coarse corn meal. Add salt, baking powder and sugar. Mix again.
* In another bowl combine the eggs and milk.
* Mix the above wet and dry ingredients together into a dough consistency.
* Wet hands with cold water and put the dough evenly into the bottom of the pan.
* For filling: slice the peaches into wedges and stand the wedges vertically in the dough, covering the entire surface of the dough.
* For topping: combine the sugar and cinnamon and sprinkle all over the peaches until heavily covered.
* Dot the 1/2 stick butter over the sugar and cinnamon mixture.
* Combine 3/4 to 1 pint of heavy cream with the 2 eggs and 1 yolk and mix thoroughly, spoon over entire cake.
* Bake in 400 degree oven for 50-60 minutes. When tested with a toothpick, dough should be dry – custard topping will remain wet.

Progressive, and at times valiant. A lot of my perspective comes from my experience in these times. This year I find myself at a very good place in my life.

I consider Judy and me very lucky. We've been soulmates for our entire adult years. We raised healthy, wonderful children who have managed to survive our parenting. Family means a lot to us and our grandchildren are a joy. We have very special friends.

I've been fortunate enough to earn a good living, and we have three residences we love that are exciting to arrive at and difficult to leave. We have traveled the world and visited exotic places with many of our friends and family. We give good parties, and have celebrated and entertained many occasions over the years with friends and family. We've fulfilled our passion for collecting. In general, we live and appreciate the good life.

I've had a successful business career where I achieved a sense of accomplishment and fulfilled most of my creative needs. For the most part, I enjoyed all of my fifty years of work. It seems like a blink of the eyes from a man's greatest success to retirement.

All our various interests and experiences in life make us who we are today. By the time you think you have the whole thing figured out, it's time to move on and let the next generation figure it out for themselves.

"L'chaim! **To life!"**

By the way, the secret of life is never taking yourself too seriously.

Epilogue

Dear Willi, Ben, Jade, Milo, Dylan, Sasha, Ruby,

When I finished this book, I realized how much of it is for you. It includes stories of family and grandparents, going back to when they first came to America—the Land of Opportunities. They worked long and hard. They left family values and a legacy that provided a strong foundation. Grandpa and then your fathers built on this foundation and now it's there for you to build on. Your opportunities are unlimited, but that doesn't come without responsibilities. When you think that school is tough, just remember that they didn't have the choice to go to school. Think about them and the examples they set with hard work, respect, and discipline. Your parents will always be there for you, just as I was there for your father and my father was always there for me. Grandma and Grandpa will always be there for you.

As you get older you realize that no one owes you anything. But you do owe something to yourself, to your family and to your country: to live good and decent and meaningful lives.

When I was young, a teacher taught me something that Rudyard Kipling wrote, "As you go through life you will be well served to take six wise men with you; who, what, why, where, when and how." As you get older you will realize how important it is to keep everything in the right perspective. Once you are able to realize what's important, you will never sweat life's hiccups.

You must mean what you say and say what you mean. Always be sincere. As you go through life you have to be genuine to yourself. I'm proud of you today and I am sure that I will always be proud of you. Study, have fun and do what you know is right. Pursue all of your dreams; most will be reachable.

I love you very much,

Grandpa

Acknowledgments

I am both lucky and grateful to have Debbie and Mitchell, my daughter-in-law and son, arrange this opportunity for me to put my life's experiences and life's lessons into a book. Debbie's gave me constant encouragement early on to write it—even though being a first time writer—I was unaware that she had me on a record-setting deadline of seven months when most books take 3 to 7 years. And don't worry Debbie, I finally did get rid of the rash!

Thanks to Kate Winters, manager and chief editor of Biography for Everyone, LLC and her team on completing this project in record time. Kate was invaluable in suggesting what stories were not necessary and what stories might be better used in different chapters. Without her knowledge and feel, this book might have been twice as long and not as good. Special thanks to Margaret Fraser, writer, in helping me restructure and at the same time capture the stories as I told them, while making my changes on numerous rewrites; Caren Hackman, graphic artist and designer, for her patience and talent in spending hours arranging the photographs with me; and the transcriptionists and copy editors who made this book possible.

Thanks to Judy, who sacrificed going out many evenings and weekends and who suffered through reading the entire original manuscript—as well as reading several revisions and making countless valuable changes and comments. Any more of *Reunion Tour*, I am afraid that Judy would have gone on tour!

A special thanks to Roger, my brother and partner, for close to 50 years, who helped jog my memory of our many years and times together and for always being there for me.

Thank you to Patty Fleishman, who typed and retyped, cut and pasted, Xeroxed and faxed, then changed and rechanged most of the pages. Her help was invaluable.

Several people have read parts of the manuscript. I would like to thank Stanley Rabinowitz and Ken Goldman for bringing the diplomat out in me. Thanks to my son Mitchell and nephew Scott for their input on Reckson history, and along with Gregg dubbed my life lessons "Donaldisms." Thanks also to Carol Allen and to my daughter-in-law, Jackie, for their suggestions.

I'm lucky to have had Bob Yaffe as a friend who chronicled so many of our trips and celebrations with great photos, many of which appear in this book. I'm thankful to my son Mark for helping me sort through hundreds of photos.

I am grateful to have numerous friends who refreshed my memory of our good times together. Thanks to Eddie Blumenfeld, Iris Rabinowitz and Arlene Karyo.

A special thanks to all of my friends and family who appear in this book. Without them the book wouldn't have been possible or necessary.

Family tree

- GLORIA (OHRBACH) RECHLER — mother (1912)
- WILLIAM RECHLER — father (1911)

DONALD RECHLER (1934 –)

- MINNIE (KATZEN) GOTTLIEB — mother (1898)
- REUBEN GOTTLIEB — father (1905)

JUDY GOTTLIEB RECHLER (1938 –)

MITCHELL RECHLER (1959) marries DEBBIE ULLIAN
- WILLI (1990)
- BENJAMIN (1992)

GLENN RECHLER (1962) marries TRACEY WEAVER
- MILO (1996)
- SASHA (2000)
- RUBY (2000)

MARK RECHLER (1966) marries JACKIE GEWIRTZ
- DYLAN (1998)
- JADE (1995)